Biochemistry

Review & Assessment

(MCQs, Clinical Case Studies, Viva/Short Questions)

Dr. U. Satyanarayana
Dr. U. Chakrapani

Biochemistry

Review & Assessment

(MCQs, Clinical Case Studies, Viva/Short Questions)

Dr. U. Satyanarayana

M.Sc., Ph.D., F.I.C., F.A.C.B.

Professor of Biochemistry & Director (Research)
Dr. Pinnamaneni Siddhartha Institute of Medical Sciences
(Dr. NTR University of Health Sciences)
Chinaoutpalli, Gannavaram (Mdl)
Krishna (Dist), A.P., India

Dr. U. Chakrapani

M.B.B.S., M.S., D.N.B.

Biochemistry Review & Assessment
(MCQs, Clinical Case Studies, Viva/Short Questions)
Satyanarayana and Chakrapani

ELSEVIER

RELX India Pvt. Ltd.

Published by RELX India Pvt. Ltd., under special arrangement with Books & Allied (P) Ltd.

First Published: September 2023

ISBN: 978-81-312-6943-5

e-Book ISBN: 978-81-312-6944-2

This book is part of the Set ISBN: 978-81-312-6941-1. Not to be sold separately.

Medical knowledge is constantly changing. As new information becomes available, changes in treatment, procedures, equipment and the use of drugs become necessary. The author, editors, contributors and the publisher have, as far as it is possible, taken care to ensure that the information given in this text is accurate and up-to-date. However, readers are strongly advised to confirm that the information, especially with regard to drug dose/usage, complies with current legislation and standards of practice. *Please consult full prescribing information before issuing prescriptions for any product mentioned in this publication.*

Biochemistry Review & Assessment, by Dr. U. Satyanarayana and Dr. U. Chakrapani is co-published by an arrangement between RELX India Private Limited and Books & Allied (P) Ltd.

RELX India Pvt. Ltd.

Registered Office: 818, 8th Floor Indraprakash Building, 21, Barakhamba Road, New Delhi-110001.

Corporate Office: 14th Floor, Building No. 10B, DLF Cyber City, Phase II, Gurgaon-122002, Haryana, India.

BOOKS & ALLIED (P) Ltd.

Registered Office: No.1E(1) 'Shubham Plaza', 83/1 Dr. S.C. Banerjee Road, Kolkata-700010, West Bengal, India.

Cover Design : Depicts the universal energy currency of the living world—ATP, predominantly synthesized by the mitochondria of the cell (the functional unit of life), in comparison with the international currencies—$, £, €, ₹, ¥.

Authors sponsored and supported by ***Uppala Author-Publisher Interlinks***, Vijayawada-520008

Typeset at Books & Allied (P) Ltd., Kolkata-700010

Printed and bound in India at Sanat Printers

Reviewers and Contributors of the Supplement to Biochemistry

We gratefully acknowledge the following faculty members who have critically reviewed and revised this supplement to Biochemistry, containing MCQs, Clinical Case Studies and Viva/Short Questions.

Dr. E. Suresh, MD (Biochemistry)

Associate Professor of Biochemistry
Dr. Pinnamaneni Siddhartha Institute of Medical Sciences
Chinaoutpalli, A.P.

Dr. (Ms) U. B. Vijaya Lakshmi, MD (Biochemistry)

Professor of Biochemistry
N.R.I. Medical College
Chinakakani, Guntur, A.P.

Dr. (Ms) P. Chudamani, MD (General Medicine)

Associate Professor of General Medicine
N.R.I. Medical College
Chinakakani, Guntur, A.P.

The contributions of MD (Biochemistry) and MD (General Medicine) Professors are unique and significant, for perfect medical/clinical orientation of MCQs, Clinical Case Studies and Viva/Short Questions in Biochemistry subject. This fully satisfies NMC guidelines of Competency-Based Medical Education (CBME) Curriculum for assessment and evalution, besides helping MBBS students to boldly and confidently face the examinations.

Dr. U. SATYANARAYANA
Dr. U. CHAKRAPANI

Preface

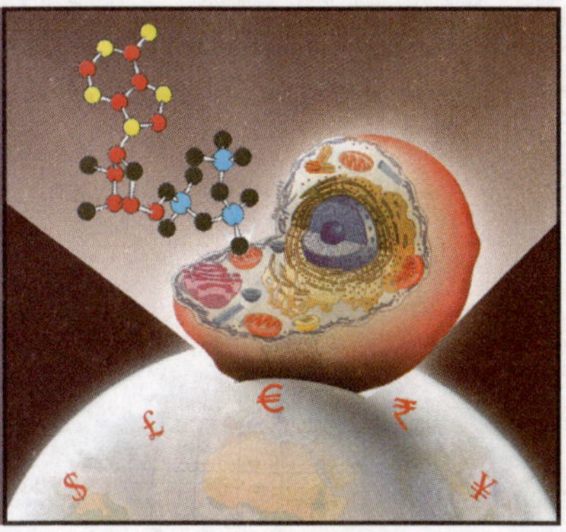

This supplement, ***Biochemistry Review & Assessment***, is intended to be the best companion to MBBS students and all others learning Biochemistry. It is organized on a sound pedagogical basis — of the Socratic method of constant questioning and finding precise answers. Thanks to the method, the 'mystery' of Biochemistry unravels and it becomes understandable, assimilable and reproducible. This review is prepared as per the Competency-Based Medical Education (CBME) curriculum of National Medical Commission (NMC).

This supplement was thoroughly ***reviewed*** and contributed by two ***MD*** (***Biochemistry***) ***Professors*** and one ***MD*** (***General Medicine***) ***Professor*** for perfect medical/clinical orientation of the subject. The MCQs, Clinical Case Studies and Viva/Short Questions, arranged chapterwise, will equip the students with sound knowledge, for assessment and evaluation, besides boldly facing any kind of examination.

Multiple Choice Questions (MCQs)

The MCQ is regarded as the most reliable method for an indepth assessment of a subject knowledge. In fact, they now form part of evaluation process as per NMC recommendations and the same is being followed by the universities. Each one of the MCQs (1200+) is 4-answer-choice category with well-chosen distractors. A novel freature of this supplement is some clinical case-related MCQs which are highly useful for medical students. All the MCQs have been framed chapter-wise for a quick review of the subject and to help the students to face examinations confidently. These include university examinations, National Exit Test (NExT) of NMC, entrance examinations for PG courses, United States Medical Licensing Examination (USMLE) etc.

Clinical Case Studies

The ultimate goal of the discipline of Biochemistry is laboratory application for the diagnosis and prognosis of diseases. And it is no exaggeration that every disease has an underlying biochemical cause! Clinical Cases (100+) covering a wide range of disorders are described in the supplement, along with the biochemical data, followed by relevant structured questions (in the same pattern of theory examination) and answers.

The cases are made as simple as possible for easy understanding. The clinical histories and laboratory data are very briefly given from the Biochemistry perspective (Note : In normal clinical practice, investigative data from microbiology, pathology, haematology, etc., are also taken into consideration while making a diagnosis. The clinical case studies described however, confine themselves to subject of Biochemistry, for practical reasons). Further, the clinical cases have been so described as to meet the academic interest and relevance of the Biochemistry subject. And for this reason, some less common disorders, and advanced laboratory investigative data also find a place while dealing with some cases.

The clinical case studies (100+) will help for a comprehensive understanding of the subject of Biochemistry with particular reference to clinical/diagnostic application. In fact, clinical cases have now become a part of theory examination in almost all universities. Therefore, this supplement is a boon to MBBS students.

Viva/Short Questions

The Viva/Short Questions (800+) have been so framed as to simplify the subject in order to help the students learn the subject continuously and logically and face the oral and theory (short notes) examinations with confidence. They will also be effective in two significant respects : in self-assisted learning and in making a rapid review of the subject.

Dr. U. SATYANARAYANA

Dr. U. CHAKRAPANI

Acknowledgements

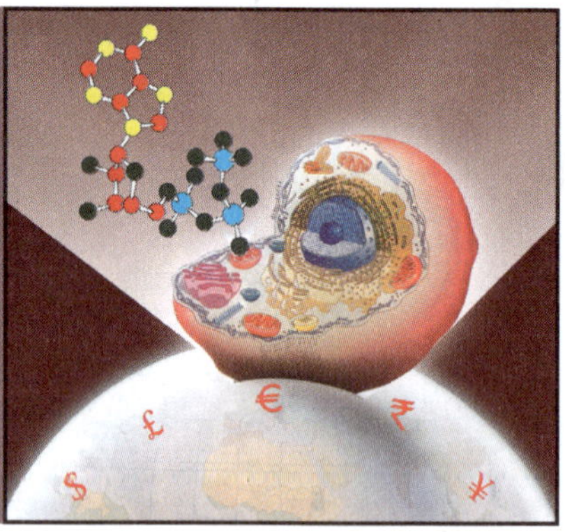

I owe a deep debt of gratitude to my parents, the late Sri U. Venkata Subbaiah and Smt. Vajramma, for cultivating in me the habit of early rising. This supplement Biochemistry Review & Assessment was written in the early morning hours (between 2 a.m-6 a.m.), when the world around was fast asleep!

I am highly grateful to the initiative, interest, enthusiasm, and dynamism of (late) Mr Arunabha Sen (Director, Books and Allied Pvt. Ltd., Kolkata) who was instrumental in publishing all my five books. We had 20 years of long and fruitful association. The sudden and premature demise of Mr Arunabha Sen (on 07-09-2018) remains a heavy professional and personal loss to me.

I thank Dr. E. Suresh, MD, Associate Professor of Biochemistry, Dr. PSIMS & RF, Chinaoutpalli, A.P. who has thoroughly reviewed and significantly contributed for the preparation of MCQs, Clinical Case Studies, Viva/Short questions, with special emphasis on medical/clinical orientation of Biochemistry. I acknowledge the contributions of Dr (Ms) U.B. Vijaya Lakshmi, MD, Professor of Biochemistry, NRI Medical College, Guntur, AP, who has helped to improve the quality of Biochemistry Review & Assessment. I am grateful to Dr. (Ms) E. Chudamani, MD, Associate Professor of General Medicine, NRI Medical College, Guntur for her significant contribution to prepare Clinical Case Studies, from a clinician's point of view.

I thank Ms Ruba Sen, Managing Director and Mr Depayan Sen, Director, Books and Allied Pvt. Ltd. for their support and encouragement to bring out the Biochemistry Review. I am indebted to Mr Shyamal Bhattacharya for his excellent page making. I am grateful to Mr Abhijit Ghosal for his help in cover design.

I express my gratitude to the officials of Elsevier (RELX India Pvt. Ltd.) for their constant support and encouragement – Mr Shanker Kaul, Managing Director; Mr Mainak Chattopadhyay – Director Sales and Channel Management, Mr Vidur Renjen, Director Education; Ms Nimisha Goswami, Director Content Strategy – India/SEA; Ms Debyani Roy, Regional Manager IPR; Ms Shabina Nasim, Head of Content Project Management – APAC; Mr Arvind Kaul, Content Strategist.

I thank my wife Krishna Kumari, my younger son Amrut Pani and my daughters-in-law Hima Bindu and Oohasri for their constant encouragement. My special thanks to my grand daughter Maahe (12 years), and grandsons Dev Sai (6 years) and Rishva (<1 year), whose eversmiling faces, sweet words and deeds infuse energy into my academic activities. I am grateful to Uppala Author-Publisher Interlinks, Vijayawada for sponsoring and supporting me to bring out this supplement to Biochemistry.

Dr. U. SATYANARAYANA

Contents

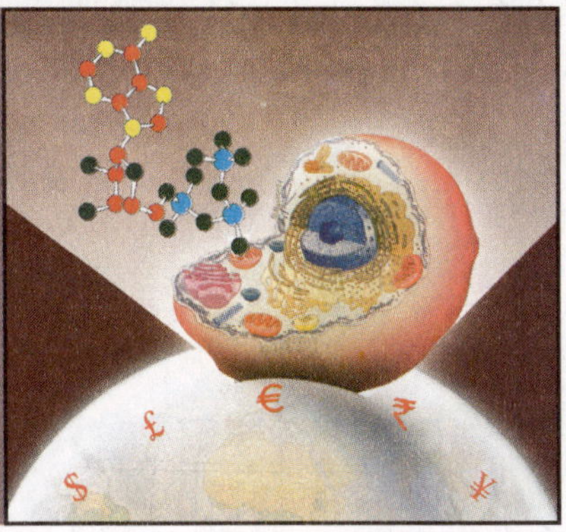

I Chemical Constituents of Life

1. Biomolecules and the cell 1
2. Carbohydrates 2
3. Lipids 4
4. Proteins and amino acids 6
5. Nucleic acids and nucleotides 10
6. Enzymes 12
7. Vitamins 18

II Physiological Biochemistry

8. Digestion and absorption 30
9. Plasma proteins 32
10. Hemoglobin and porphyrins 35
11. Biological oxidation 39

III Metabolisms

12. Introduction to metabolism 41
13. Metabolism of carbohydrates 42
14. Metabolism of lipids 49
15. Metabolism of amino acids 58
16. Integration of metabolism 70
17. Metabolism of nucleotides 72
18. Mineral metabolism 75

IV Clinical Biochemistry and Nutrition

19. Hormones 84
20. Organ function tests 89
21. Water, electrolyte and acid-base balance 94
22. Tissue proteins and body fluids 99
23. Nutrition 100

V Molecular Biology and Biotechnology

24. DNA-replication, recombination and repair 106
25. Transcription and translation 107
26. Regulation of gene expression 111
27. Recombinant DNA and biotechnology 113

VI Current Topics

28. Human genome project 115
29. Gene therapy 115
31. Metabolism of xenobiotics (detoxification) 116
32. Prostaglandins and related compounds 117
33. Biological membranes and transport 118
34. Free radicals and antioxidants 120
35. Environmental biochemistry 120
36. Insulin, glucose homeostasis, and diabetes mellitus 121
37. Cancer 127
38. Acquired immunodeficiency syndrome (AIDS) 129

CHEMICAL CONSTITUENTS OF LIFE I

1. Biomolecules and the Cell

MCQs

1. **The most predominant chemical constituent of life**
   (A) Water (B) Protein
   (C) Lipid (D) Carbohydrate.
2. **One of the following cannot form a polymer in the living cell**
   (A) Monosaccharides (B) Amino acids
   (C) Fatty acids (D) Ribonucleotides.
3. **The mitochondrial DNA (mtDNA) is**
   (A) Circular double stranded
   (B) Circular single stranded
   (C) Linear double helix
   (D) None of the above.
4. **The cellular organelle regarded as the digestive tract of the cell**
   (A) Nucleus (B) Golgi appratus
   (C) Peroxisome (D) Lysosome.
5. **The cellular organelle predominantly responsible for respiration and energy metabolism**
   (A) Glyoxysomes (B) Golgi apparatus
   (C) Mitochondria (D) Nucleus.
6. **The pH of the lysosomal matrix is**
   (A) Equal to that of cytosol
   (B) More acidic than cytosol
   (C) More basic than cytosol
   (D) Neutral

Answers

1. **(A)** 2. **(B)** 3. **(A)** 4.**(D)** 5. **(C)** 6. **(B)**.

Clinical Case Study 1.1 (Source : Biochemistry, Page 6)

A child of 2 years age had delayed growth, short height, small face, hair toss. It was observed that the child was aging faster than usual.

1. **What is the provisional diagnosis?**
2. **What is the biochemical basis of this disorder?**
3. **What are the complications of this disease?**

Ans. :

1. **Hutchinson-Gilford progeria syndrome (HGPS) or simply progeria.**
2. A gene mutation is responsible for progeria. Normally, Lamin A (LMNA) gene produces protein Lamin A that holds nucleus at the center of the cell and maintains the integrity of the cell. Mutation of lamin A gene results in an abnormal protein, progerin that makes cells unstable leading to premature aging.
3. Alopecia (hair loss), wrinkled skin, small face, loss of eyesight, kidney failure, atherosclerosis, heart disease. Average life span is around 15 years.

For related clinical case study, refer 14.4

VIVA/Short Questions

1. **Name the structural and functional unit of life.**

The cell.

2. **Which element is most abundantly present in biomolecules?**

Carbon.

3. **The hydrolase enzymes of the cell are predominantly localized in the subcellular organelle, namely**

Lysosomes.

4. **Name the pigment that accumulates in the cells as the age advances.**

Lipofuscin.

5. **Where are the enzymes of energy metabolism present in a prokaryotic cell?**

Cell membrane.

6. **Name the disorder where rapid aging begins at birth**

Hutchinson-Gilford progeria syndrome (HGPS).

7. **What do you understand by apoptosis?**

Programmed cell death or cell suicide.

8. **What is the disorder due to the absence of certain hydrolases in lysosomes?**

Inclusion cell (I-cell) disease.

2. Carbohydrates

MCQs

1. **One of the following, although a genuine carbohydrate, does not satisfy the empirical formula $(C.H_2O)_n$ of carbohydrates**
   (A) Deoxyribose (B) Ribose
   (C) Mannose (D) Xylulose.
2. **All of the following are aldoses except**
   (A) Erythrose (B) Ribose
   (C) Ribulose (D) Glucose.
3. **All of the following are ketoses except**
   (A) Ribulose (B) Fructose
   (C) Erythrose (D) Sedoheptulose.
4. **The trisaccharide-raffinose is composed of**
   (A) Fructose, galactose and glucose
   (B) Fructose, mannose and glucose
   (C) Fructose, fructose and glucose
   (D) Fructose, galactose and mannose.
5. **All of the following polysaccharides contain glucose except**
   (A) Starch (B) Glycogen
   (C) Inulin (D) Cellulose.
6. **The following monosaccharides contain 6 carbons except**
   (A) Glucose (B) Mannose
   (C) Galactose (D) Xylulose.
7. **One of the following monosaccharides is a tetrose (contains 4 carbons)**
   (A) Ribulose (B) Xylulose
   (C) Deoxyribose (D) Erythrose.
8. **The naturally occurring monosaccharides in the mammalian tissues are of**
   (A) D-configuration (B) L-configuration
   (C) Racemic mixture (D) DL-mixture.
9. **The number of asymmetric carbons present in glucose**
   (A) 2 (B) 4 (C) 6 (D) 8.
10. **The reducing property of carbohydrates is attributed to**
    (A) Free aldehyde or keto group
    (B) Free hydroxyl or amino group
    (C) Free carboxyl or sulfhydryl group
    (D) Free hydroxyl and carboxyl group.
11. **The oxidation of terminal alcohol group ($CH_2OH \longrightarrow COOH$) of glucose results in the formation of**
    (A) Gulonic acid (B) Glucuronic acid
    (C) Sorbitol (D) Gulonolactone.
12. **Glucose on reduction results in the formation of the following alcohol**
    (A) Mannitol (B) Ribitol
    (C) Dulcitol (D) Sorbitol.
13. **Reducing sugars form osazones with phenylhydrazine. The two carbon atoms involved in osazone formation**
    (A) 1 and 2 (B) 3 and 4
    (C) 4 and 5 (D) 5 and 6.
14. **Needle shaped osazones are formed by**
    (A) Glucose (B) Fructose
    (C) Mannose (D) All of the above.
15. **One of the following is a non-reducing disaccharide**
    (A) Lactose (B) Maltose
    (C) Trehalose (D) Isomaltose.
16. **In sucrose, α-D glucose and β-D fructose are held together by a glycosidic bond at**
    (A) C_1 of glucose and C_1 of fructose
    (B) C_1 of glucose and C_2 of fructose
    (C) C_1 glucose and C_3 of fructose
    (D) C_2 of glucose and C_2 of fructose.
17. **All of the following can form osazones except**
    (A) Glucose (B) Lactose
    (C) Maltose (D) Sucrose.
18. **The disaccharide lactose is composed of**
    (A) Glucose and galactose
    (B) Glucose and mannose
    (C) Glucose and glucose
    (D) Glucose and fructose.
19. **One of the following can exist as linear and branched polymers**
    (A) Proteins (B) Nucleic acids
    (C) Polysaccharides (D) Lipids.
20. **All of the following are polymers of glucose except**
    (A) Starch (B) Inulin
    (C) Dextrin (D) Cellulose.
21. **One of the following polysaccharides on hydrolysis forms β-D glucose**
    (A) Cellulose (B) Starch
    (C) Glycogen (D) Inulin.

22. One of the following glycosaminoglycans acts as an anticoagulant

(A) Keratan sulfate (B) Hyaluronic acid
(C) Heparin (D) Chondroitin sulfate.

23. The following monosaccharide is predominantly utilized by sperms for energy

(A) Glucose (B) Fructose
(C) Galactose (D) Mannose.

24. Heparin is composed of the following as repeating units

(A) D-glucuronic acid and N-acetyl glucosamine
(B) D-glucuronic acid N-acetyl galactosamine 4-sulfate
(C) D-glucuronate 2 sulfate and N-sulfoglucosamine 6-sulfate
(D) L-iduronic acid and N-acetyl glucosamine 4-sulfate.

Answers

1. **(A)** 2. **(C)** 3. **(C)** 4. **(A)** 5. **(C)** 6. **(D)** 7. **(D)**. 8. **(A)** 9. **(B)** 10. **(A)** 11. **(B)** 12. **(D)** 13. **(A)** 14. **(D)** 15. **(C)** 16. **(B)** 17. **(D)** 18. **(A)** 19. **(C)** 20. **(B)** 21. **(A)** 22. **(C)** 23. **(B)** 24. **(C)**.

For clinical case studies related to Carbohydrate Metabolism, refer Chapter 13, and Chapter 36 for clinical case studies of Diabetes Mellitus.

VIVA/Short Questions

1. Define carbohydrates.

Polyhydroxy aldehydes or ketones, or compounds which produce them on hydrolysis.

2. What do you understand by glycome and glycomics?

Glycome represents the entire spectrum of carbohydrates of an organism. Glycomics is the study of glycomes.

3. What do you understand by monosaccharides?

The simplest group of carbohydrates with a general formula $C_n(H_2O)_n$. e.g., glucose is aldohexose while fructose is a ketohexose.

4. What are disaccharides?

Composed of two molecules of the same or different monosaccharides e.g., sucrose (glucose + fructose), lactose (glucose + galactose).

5. Define polysaccharides.

Polymers of monosaccharides, homopolysaccharides (e.g., starch, inulin, cellulose) and heteropolysaccharides with two or more different types of monosaccharides (e.g., heparin, chondroitin sulfate).

6. Name the compound that has been chosen as the reference for writing the structures of carbohydrates.

Glyceraldehyde.

7. What are epimers?

Two monosaccharides differ from each other in their configuration around a single carbon atom. e.g., glucose and galactose are C_4-epimers; glucose and mannose are C_2 epimers.

8. What are anomers?

The α- and β-cyclic forms of D-glucose are known as anomers. They differ from each other only around carbon 1 (anomeric carbon).

9. What do you understand by mutarotation?

The change in the specific optical rotation representing the interconversion of α- and β-anomers of D-glucose to an equilibrium mixture.

10. What is tautomerization?

The process of shifting a hydrogen atom from one carbon to another to produce enediols.

11. What do you understand by glycosides?

Compounds formed when the hydroxyl group of anomeric carbon of carbohydrates reacts with a hydroxyl group of another carbohydrate or a non-carbohydrate.

12. What is aglycone?

Aglycone refers to the non-carbohydrate moiety found in glycosides.

13. Name the vitamin whose structure closely resembles that of a monosaccharide.

Ascorbic acid (vitamin C).

14. What is the carbohydrate of importance for the nutrition of young mammals?

Lactose (milk sugar).

15. Name the group of polysaccharides which on hydrolysis yield a single type of monosaccharides?

Homopolysaccharides (homoglycans) e.g., starch, inulin.

16. What is starch?

Polysaccharide composed of D-glucose units held together by α-glycosidic bonds, (α 1→ 4 linkages; at branching points α 1→ 6 linkages).

17. What is inulin and what is its importance?

Inulin is a polymer of fructose (fructosan). It is useful for assessing kidney function through the measurement of glomerular filtration rate (GFR).

18. What do you understand by animal starch?

Glycogen, the carbohydrate reserve of animals, is frequently referred to as animal starch.

19. Why humans cannot digest cellulose?

Due to lack of the enzyme that cleaves β-glycosidic bonds.

20. What are mucopolysaccharides?

Mucopolysaccharides (commonly known as glycosaminoglycans) are heteropolysaccharides composed of sugar derivatives (mainly amino sugars and uronic acids).

21. What is the importance of hyaluronic acid?

Serves as lubricant and shock absorbant in joints.

22. What are glycoproteins?

The proteins structurally associated with the carbohydrates.

23. Name some diseases associated with carbohydrates.

Diabetes mellitus, glycogen storage diseases, galactosemia.

3. Lipids

MCQs

1. The characteristic features of lipids include their

(A) Solubility in organic solvents
(B) Insolubility in water
(C) Actual or potential relation to fatty acids
(D) Characters listed in A, B and C.

2. Waxes contain one of the following alcohols

(A) Glycerol (B) Cetyl alcohol
(C) Ethyl alcohol (D) Sorbitol.

3. Most of the naturally occurring lipids contain fatty acids with

(A) Even number of carbons
(B) Odd number of carbons
(C) Even and odd number of carbons equally
(D) None of the above.

4. All of the following are saturated fatty acids except

(A) Oleic acid (B) Arachidic acid
(C) Stearic acid (D) Palmitic acid.

5. All of the following are unsaturated fatty acids except

(A) Oleic acid (B) Arachidonic acid
(C) Linolenic acid (D) Myristic acid.

6. The number of carbon atoms present in stearic acid

(A) 14 (B) 16 (C) 18 (D) 20.

7. The number double bonds present in linoleic acid

(A) 1 (B) 2 (C) 3 (D) 4.

8. The following is a shorthand representation of arachidonic acid

(A) 20 : 4; 5, 8, 11, 14
(B) 18 : 3; 9, 12, 15
(C) 18 : 2; 9, 12 (D) 16 : 1; 1.

9. The unsaturated fatty acids that are essential to humans

(A) Oleic acid and linoleic acid
(B) Linoleic acid linolenic acid
(C) Linolenic acid and arachidonic acid
(D) Arachidonic acid and palmitoleic acid.

10. The naturally occurring unsaturated fatty acids exist as

(A) *cis*-isomers (B) *trans*-isomers
(C) *cis*- and *trans*-isomers
(D) None of the above.

11. One of the following lipids is not a structural component of biological membranes

(A) Phospholipids (B) Cholesterol
(C) Glycolipids (D) Triacylglycerols.

12. The triacylglycerols of plants in general contain

(A) Equal concentrations of saturated and unsaturated fatty acids
(B) Higher concentration of saturated fatty acids
(C) Higher concentration of unsaturated fatty acids
(D) Unsaturated fatty acids only.

13. The number of grams of iodine absorbed by 100 g of fat or oil is referred to as

(A) Saponification number
(B) Acid number (C) Reichert-Meissl number
(D) Iodine number.

14. Phospholipids may be regarded as the derivatives of

(A) Lecithin (B) Cephalin
(C) Phosphatidic acid (D) Plasmalogen.

15. The nitrogenous base present in lecithins

(A) Choline (B) Ethanolamine

(C) Serine (D) Inositol.

16. The absence of the following phospholipid causes respiratory distress syndrome in infants

(A) Cardiolipin (B) Dipalmitoyl lecithin

(C) Cephalins (D) Plasmalogens.

17. The action of certain hormones is mediated through the following phospholipid

(A) Phosphatidyl serine

(B) Phosphatidyl inositol

(C) Phosphatidyl choline

(D) Cephalins.

18. The following phospholipid possesses antigenic properties

(A) Cardiolipin (B) Lecithin

(C) Cephalin (D) Plasmalogen.

19. The following fatty acid liberated from membrane serves as the precursor for the synthesis of prostaglandins

(A) Oleic acid (B) Linoleic acid

(C) Linolenic acid (D) Arachidonic acid.

20. The following compound lipids contain N-acetylneuraminic acid (NANA)

(A) Phospholipids (B) Lipoproteins

(C) Gangliosides (D) Sphigomyelins.

21. One of the following is a plant sterol

(A) Cholesterol (B) Ergosterol

(C) Cortisol (D) Lanosterol.

22. All of the following are amphipathic lipids except

(A) Triacylglycerols (B) Fatty acids

(C) Phospholipids (D) Cholesterol.

Answers

1. **(D)** 2. **(B)** 3. **(A)** 4. **(A)** 5. **(D)** 6. **(C)** 7. **(B)** 8.**(A)** 9. **(B)** 10. **(A)** 11. **(D)** 12. **(C)** 13. **(D)** 14. **(C)** 15. **(A)** 16. **(B)** 17. **(B)** 18. **(A)** 19. **(D)** 20. **(C)** 21. **(B)** 22. **(A)**.

For clinical case studies related to Lipid Metabollism refer Chapter 14, and Chapter 36 for clinical case studies of Diabetes Mellitus.

VIVA/Short Questions

1. What are lipids?

Organic substances relatively insoluble in water, soluble in organic solvents (alcohol, ether etc.), actually or potentially related to fatty acids and utilized by the living cells.

2. How do you classify lipids?

Broadly classified into simple, complex (compound), derived and miscellaneous.

3. What is the difference between fats and oils?

Fats are solids while oils are liquids at room temperature.

4. What are waxes?

Esters of fatty acids (usually long chain) with alcohols other than glycerol.

5. Give some examples of compound (complex) lipids.

Phospholipids, glycolipids, lipoproteins, sulfolipids.

6. What are the main functions of lipids?

Concentrated fuel reserve, constituents of membrane structure, source of fat soluble vitamins etc.

7. What are fatty acids?

Carboxylic acids with hydrocarbons side chain e.g., palmitic acid, stearic acid.

8. What do you understand by polyunsaturated fatty acids (PUFA)?

Fatty acids containing two or more double bonds.

9. What are essential fatty acids? Name them.

The fatty acids that cannot be synthesized by the body and therefore should be supplied through the diet. Linoleic acid and linolenic acid are essential.

10. What is phrynoderma?

An essential fatty acid deficiency disorder.

11. What type of isomerism is observed in unsaturated fatty acids?

Geometric or *cis-trans* isomerism.

12. What is saponification?

The hydrolysis of triacylglycerols by alkali to produce glycerol and soaps.

13. What do you understand by the term rancidity?

Deterioration of fats and oils resulting in unpleasant taste.

14. What are phospholipids?

Complex or compound lipids containing phosphoric acid, in addition to fatty acids, nitrogenous base and alcohol.

15. What is the importance of dipalmitoyl lecithin?

Prevents the adherence of inner surface of lungs due to surface tension. Absence of this phospholipid results in respiratory distress syndrome in infants.

16. What are sphingomyelins?

A group of phospholipids containing sphingosine as the alcohol (in place of glycerol in other phospholipids).

17. List out the important functions of phospholipids.

Participate in intestinal absorption of fats; essential for the synthesis of lipoproteins; required for reverse cholesterol transport.

18. What are lipoproteins? Name them.

Molecular complexes of lipids with proteins. Very low density lipoproteins (VLDL), low density lipoproteins (LDL), high density lipoproteins (HDL) and free fatty acid-albumin complexes.

19. Which is the cyclic ring present in all the steroids?

Cyclopentanoperhydrophenanthrene (CPPP).

20. What are the important functions of cholesterol?

Involved in membrane structure and function, in the synthesis of bile acids, hormones (sex and cortical) and vitamin D.

21. Name some laboratory tests used for the detection of cholesterol.

Salkowski's test, Lieberman-Burchard reaction, ZaK's test.

22. Give the names of some important plant sterols.

Ergosterol, stigmasterol, β-sitosterol.

23. What are amphipathic lipids?

Lipids which possess both hydrophobic and hydrophilic groups.

24. Name some important amphipathic lipids.

Phospholipids, fatty acids, bile salts, cholesterol.

25. What are liposomes?

Produced when amphipathic lipids in aqueous medium are subjected to sonification. Liposomes are used as carriers of drugs to target tissues.

26. Name some diseases associated with lipids.

Atherosclerosis, obesity.

4. Proteins and Amino Acids

MCQs

1. Protein are the polymers of

(A) L-α-amino acids (B) D-α-amino acids
(C) L-β-amino acids (D) D-β-amino acids.

2. Kjeldahl's method in the laboratory employs the estimation of the following element in the biological fluids for the quantitation of proteins

(A) Carbon (B) Hydrogen
(C) Oxygen (D) Nitrogen.

3. Nitrogen is an essential component of proteins. The content of nitrogen in protein is around

(A) 8% (B) 12% (C) 16% (D) 20%.

4. There are about 300 amino acids in nature but only 20 of these are repeatedly found in the proteins isolated from different forms of life (animals, plants, microorganisms). This is because of the

(A) Universal nature of the genetic code for 20 amino acids
(B) Structural similarity between the 20 amino acids
(C) Similarities in ribosomes, the factories of protein synthesis
(D) Comparable structures of DNA.

5. One of the following amino acids does not exhibit optical isomerism

(A) Isoleucine (B) Lysine
(C) Glycine (D) Threonine.

6. The following three amino acids are branched chain

(A) Glycine, alanine, valine
(B) Valine, leucine, isoleucine
(C) Leucine, isoleucine, lysine
(D) Isoleucine, lysine, proline.

7. One of the following is an aromatic hydroxyl group containing amino acid

(A) Serine (B) Threonine
(C) Tyrosine (D) Phenylalanine.

8. All of the following are basic amino acids except

(A) Lysine (B) Arginine
(C) Histidine (D) Glutamate.

9. One of the following is an aromatic non-essential amino acid

(A) Histidine (B) Phenylalanine

(C) Tyrosine (D) Tryptophan.

10. The amino acids respectively containing guanidino, imidazole and indole groups

(A) Arginine, histidine, tryptophan

(B) Histidine, tryptophan, arginine

(C) Tryptophan, arginine, histidine

(D) Arginine, proline, tryptophan.

11. The amino acids respectively containing sulfhydryl, disulfide and thioether groups

(A) Methionine, cysteine, cystine

(B) Cysteine, cystine, methionine

(C) Methionine, cystine, cysteine

(D) Cystine, cysteine, methionine.

12. One of the following is an imino acid

(A) Histidine (B) Asparagine

(C) Tyrosine (D) Proline.

13. All of the following are essential amino acids except

(A) Valine, leucine, isoleucine

(B) Lysine, methionine, phenylalanine

(C) Tyrosine, cysteine, alanine

(D) Arginine, threonine, tryptophan.

14. The following two amino acids are semiessential

(A) Arginine, histidine

(B) Cysteine, tyrosine

(C) Proline, phenylalanine

(D) Serine, threonine.

15. The following are aromatic essential amino acids

(A) Histidine, tyrosine

(B) Phenylalanine, tryptophan

(C) Proline, histidine

(D) Tyrosine, proline.

16. All of the following are non-polar amino acids except

(A) Alanine, leucine, isoleucine

(B) Valine, methionine, phenylalanine

(C) Tryptophan, proline, alanine

(D) Serine, threonine, glutamine.

17. The following are polar amino acids except

(A) Glycine, serine, threonine

(B) Cystine, glutamine, asparagine

(C) Alanine, leucine, isoleucine

(D) Tyrosine, threonine, glutamine.

18. All of the following are non-essential amino acids except

(A) Glycine, alanine, serine

(B) Valine, methionine, tryptophan

(C) Cysteine, asparagine, glutamine

(D) Tyrosine, proline, aspartate.

19. At isoelectric pH (pI), amino acids exist as

(A) Cations (B) Anions

(C) Zwitter ions (D) Acidic ions.

20. The most abundant protein in mammals

(A) Albumin (B) Globulin

(C) Hemoglobin (D) Collagen.

21. The following amino acid derivative is found in certain blood clotting factors

(A) Hydroxyproline

(B) Hydroxylysine

(C) γ-carboxyglutamate

(D) Methylhistidine.

22. Collagen is rich in the amino acid derivative

(A) Hydroxyproline (B) γ-carboxyglutamate

(C) Methylhistidine (D) Methyllysine.

23. One of the following is a non-α-amino acid

(A) Ornithine (B) Homoserine

(C) Taurine (D) Homocysteine.

24. All of the following are non-protein amino acids except

(A) Ornithine, citrulline

(B) Thyroxine, triiodothyronine

(C) Homocysteine, homoserine

(D) Glutamine, asparagine.

25. The bacterial cell walls are rich in

(A) L-glutamate, L-alanine

(B) D-glutamate, D-alanine

(C) D-threonine, D-serine

(D) D-aspartate, D-valine.

26. One of the following amino acids is formed by the combination of two molecules of another amino acid

(A) Ornithine (B) Citrulline

(C) Cystine (D) Methionine.

27. The two cysteine residues in cystine are held together by a

(A) Peptide bond (B) Disulfide bond

(C) Hydrogen bond (D) Hydrophobic bond.

28. A decapeptide will have

(A) 10 amino acids with 9 peptide bonds

(B) 11 amino acids with 10 peptide bonds

(C) 10 amino acids with 10 peptide bonds

(D) 9 amino acids with 10 peptide bonds.

29. The following is the other name of glutathione

(A) Glycinyl-cysteinyl-glutamate

(B) Cysteinyl-glycinyl-glutamate

(C) Glutamyl-cysteinyl-glycine

(D) Glutamyl-glycinyl-cysteine.

30. The primary structure of protein involves the determination of

(A) Bonds responsible for protein structure

(B) The spacial arrangement of polypeptide chains

(C) Amino acid composition and their sequence

(D) Three dimensional structure of protein.

31. The mixture of amino acids liberated by protein hydrolysis are most commonly separated by

(A) Electrophoresis (B) Chromatography

(C) Ultracentrifugation

(D) Immunoelectrophoresis.

32. The number of polypeptide chains in a protein can be identified by treatment with

(A) Cyanogen bromide

(B) Sanger's reagent

(C) Dansyl chloride

(D) Edman's reagent.

33. The chemical name of Edman's reagent

(A) 1-fluoro 2, 4-dinitrobenzene

(B) Phenylisothiocyanate

(C) Cyanogen bromide

(D) Dansyl chloride.

34. Sanger's reagent (1-fluoro 2,4-dinitrobenzene) is used for the identification of

(A) N-terminal amino acid

(B) C-terminal amino acid

(C) Polypeptide chains

(D) Disulfide bonds.

35. Sequenator is an automatic machine employed to determine the amino acid sequence in a polypeptide chain. Its working is based on the use of

(A) Sanger's reagent (B) Dansyl chloride

(C) Cyanogen bromide

(D) Edman's reagent.

36. The structure of α-helix is characterized by all of the following except

(A) Extensive hydrogen bonding

(B) The backbone of the structure are peptide bonds

(C) Each turn of the coiled structure has 3.6 amino acids

(D) Formation of hydrogen bonds between neighbouring polypeptide chains.

37. The structure of β-pleated sheet is

(A) Spiral with H-bonds

(B) Composed of two or more polypeptide chains bound by hydrogen bonds

(C) A fully extended single polypeptide chain

(D) Not having any one of the characteristics listed above.

38. The secondary structure of a protein is stabilized by

(A) Peptide bonds (B) Hydrogen bonds

(C) Disulfide bonds (D) Hydrophobic bonds.

39. Each amino acid in a protein on an average contributes to a molecular weight of about

(A) 55 daltons (B) 110 daltons

(C) 165 daltons (D) 220 daltons.

40. The precipitation of proteins by neutral salts such as ammonium sulfate is based on the

(A) Hydration of molecules

(B) Hydrolysis of molecules

(C) Sulfation of molecules

(D) Dehydration of molecules.

41. One of the following is commonly used as a precipitating reagent to prepare protein - free filtrate of blood in the laboratory

(A) Phosphotungstic acid
(B) Sulphosalicylic acid
(C) Tannic acid
(D) Phosphomolybdic acid.

42. The minimum number of peptide bonds needed to answer biuret test

(A) 1 (B) 2 (C) 3 (D) 4.

43. One of the following amino acids answers biuret test

(A) Tryptophan (B) Tyrosine
(C) Phenylalanine (D) Histidine.

44. Denaturation of protein is characterized by all of the following except

(A) Cleavage of peptide bonds
(B) Loss of biological activity
(C) Increase in its viscosity and decrease in surface tension
(D) Its easy digestion by gastrointestinal tract enzymes.

45. One of the following is an example of transport proteins

(A) Keratin (B) Pepsin
(C) Actin (D) Albumin.

46. All of the following are structural proteins except

(A) Keratin (B) Glutelin
(C) Elastin (D) Collagen.

47. One of the following is an example of phosphoprotein

(A) Histone (B) Casein
(C) Ceruloplasmin (D) Albumin.

48. The following are incomplete or partially incomplete proteins except

(A) Gelatin (B) Zein
(C) Egg albumin (D) Meat protein.

49. Glutathione performs the following functions except

(A) Maintaining disulfide bonds in protein structure
(B) Maintaining erythrocyte membrane structure
(C) Helping in the intestinal transport of amino acids
(D) Helping in the transport of steroid hormones.

50. The physician diagnosed a medical student as a victim of Chinese restaurant syndrome (flu-like symptoms). The following compound is responsible for this disorder.

(A) Pyrrolysine (B) Selenocysterine
(C) Monosodium glutamate
(D) Monosodium aspartate

Answers

1. **(A)** 2.**(D)** 3. **(C)** 4. **(A)** 5. **(C)** 6. **(B)** 7. **(C)** 8. **(D)** 9. **(C)**. 10. **(A)** 11. **(C)** 12. **(D)** 13. **(C)** 14. **(A)** 15. **(B)** 16. **(D)** 17. **(C)** 18. **(B)** 19. **(C)** 20. **(D)** 21. **(C)** 22. **(A)** 23. **(C)** 24. **(D)** 25. **(B)** 26. **(C)** 27. **(B)** 28. **(A)** 29. **(C)** 30. **(C)** 31. **(B)** 32. **(C)** 33. **(B)** 34. **(A)** 35. **(D)** 36. **(D)** 37. **(B)** 38. **(B)** 39. **(B)** 40. **(D)** 41. **(A)** 42. **(B)** 43. **(D)** 44. **(A)** 45. **(D)** 46. **(B)** 47. **(B)** 48. **(C)** 49. **(D)** 50. **(C)**.

For clinical case studies related to Amino Acid Metabolism, refer Chapter 15.

VIVA/Short Questions

1. What are proteins?

The polymers of L-α-amino acids.

2. What are the main functions of proteins?

Perform static (structural) and dynamic functions – proteins acting as enzymes, hormones, immunoglobulins, blood clotting factors etc.

3. List out the major elements, commonly found in protein structure.

Carbon, hydrogen, oxygen, nitrogen and sulfur.

4. What are amino acids?

Organic compounds containing two functional groups-amino ($—NH_2$) and carboxyl (—COOH).

5. What do you understand by α-amino acids?

The amino acids in which both carboxyl and amino groups are attached to the same carbon atom.

6. Name the amino acids with aliphatic side chains.

Glycine, alanine, valine, leucine and isoleucine. All these are monoamino monocarboxylic acids.

7. List out the basic amino acids.

Lysine, arginine, histidine. These are dibasic monocarboxylic acids.

8. Name the aromatic amino acids.

Phenylalanine, tyrosine and tryptophan (Note : some authors include histidine also).

9. What are essential amino acids? Name them.

The amino acids which cannot be synthesized by the body and therefore need to be supplied through the diet - **V**aline, **I**soleucine, **L**eucine, **L**ysine, **M**ethionine, **P**henylalanine, **T**hreonine, **T**ryptophan.

10. How are the amino acids classified from the metabolic point of view?

(a) ***Glycogenic*** **:** e.g., alanine, aspartate, glycine.

(b) ***Ketogenic*** **:** e.g., leucine, lysine.

(c) ***Glycogenic or ketogenic*** **:** e.g., isoleucine, phenylalanine, tryptophan.

11. What is a zwitterion?

A hybrid molecule containing positive and negative ionic groups.

12. What do you understand by isoelectric pH?

The pH at which a molecule exists as a zwitterion and therefore carries no net charge.

13. What do you mean by non-protein amino acids?

The amino acids which are never found in protein structure. e.g., ornithine, citrulline, thyroxine.

14. Give some examples of non-α-amino acids.

β-Alanine, β-aminoisobutyric acid, γ-aminobutyric acid, δ-aminolevulinic acid.

15. What is a peptide bond?

A peptide bond (—CO—NH—) is formed when an amino group of an amino acid combines with a carboxyl group of another amino acid. It is the backbone of protein structure.

16. What is Sanger's reagent?

1-fluoro 2,4-dinitrobenzene (FDNB). It is used to detect the N-terminal amino acid of a polypeptide or a protein.

17. What is Edman's reagent?

Phenylisothiocyanate. It is used to sequentially determine the N-terminal amino acids in protein structure.

18. What do you understand by sequenator?

An automatic machine employed to determine the amino acid sequence in a polypeptide.

19. What is an α-helix?

The most common secondary structure of proteins. It is stabilized by extensive hydrogen bonding. Each turn of α-helix contains 3.6 amino acids and travels a distance of 0.54 nm.

20. Name the bonds responsible for the tertiary structure of proteins.

Hydrogen bonds, disulfide bonds, electrostatic bonds and hydrophobic interactions.

21. What are the covalent bonds found in protein structure?

Peptide and disulfide bonds.

22. Name the non-covalent bonds usually found in protein structure.

Hydrogen bonds, hydrophobic bonds, electrostatic bonds and Van der Waals forces.

23. What is denaturation?

The disorganization of native protein structure, with a loss of secondary, tertiary and quaternary structure of proteins.

24. What are the characteristics of denatured protein?

Loses its helical structure, biological activity and becomes insoluble. However, the peptide bonds are not broken in denaturation.

25. How are proteins classified?

Based on function (structural protein, hormonal protein, catalytic protein) or more appropriately based on their chemical nature and solubility properties – simple, conjugated and derived.

26. Give some examples of conjugated proteins.

Nucleoproteins, lipoproteins, phosphoproteins, metaloproteins, glycoproteins.

27. How are proteins classified from the nutritional point of view?

Complete proteins, partially incomplete proteins, incomplete proteins.

28. What are the important functions of glutathione?

It maintains the sulfhydryl (—SH) groups in proteins; protects hemoglobin from getting oxidized; maintains RBC membrane structure and integrity; involved in the transport of certain amino acids in the intestine and kidney.

5. Nucleic Acids and Nucleotides

MCQs

1. One of the following is regarded as the reserve bank of genetic information in higher organisms

(A) Deoxyribonucleotide

(B) Deoxyribonucleic acid

(C) Protein (D) Ribonucleic acid.

2. DNA is found in all of the following except

(A) Nucleus (B) Mitochondria

(C) Golgi apparatus (D) Chloroplasts.

3. DNA is the polymer of

(A) Ribonucleotides

(B) Deoxyribonucleotides

(C) Deoxyriboses (D) Ribose sugars.

4. A nucleotide is composed of the following except

(A) Nitrogenous base (B) Pentose sugar

(C) Phosphate (D) Glycerol.

5. Nucleotides perform all of the following functions except they are

(A) Structural units of DNA and RNA

(B) Catalysts in function

(C) Components of certain coenzymes

(D) Regulators of metabolic reactions.

6. The difference between nucleotide and nucleoside

(A) Absence of a phosphate in nucleoside

(B) Presence of a base in nucleoside

(C) Absence of a pentose in nucleotides

(D) Absence of a base in nucleoside.

7. One of the following is not a purine

(A) Adenine (B) Uric acid

(C) Thymine (D) Guanine.

8. All of the following are pyrimidines except

(A) Cytosine (B) Thymine

(C) Uracil (D) Xanthine.

9. Deoxyribose has one oxygen less than ribose at carbon

(A) 2 (B) 3 (C) 4 (D) 5.

10. Thymine (T) and uracil (U) differ in their structure by the presence (in T) or absence (in U) of a/an

(A) Amino group (B) Methyl group

(C) Hydroxyl group (D) Phosphate group.

11. One of the following is a major purine base found in tea

(A) Caffeine (B) Theobromine

(C) Theophylline (D) Hypoxanthine.

12. All of the following bases exhibit tautomerism except

(A) Adenine (B) Guanine

(C) Cytosine (D) Thymine.

13. Chemically, uric acid is

(A) 6-oxypurine (B) 2,6-dioxypurine

(C) 2,6,8-trioxypurine (D) 6-aminopurine.

14. The purine and pentose (ribose or deoxyribose) sugar of a nucleoside are held together by the following glycosidic bond

(A) N^8-C_1 (B) N^3-C_1 (C) N^7-C_3 (D) N^7-C_5.

15. The pyrimidine and pentose sugar of a nucleoside are held together by the following glycosidic bonds

(A) N^1-C_1 (B) N^1-C_2 (C) N^3-C_1 (D) N^3-C_5

16. One of the following is absent in DNA structure

(A) Adenine (B) Guanine

(C) Cytosine (D) Uracil.

17. RNA contains all of the following except

(A) Adenine (B) Guanine

(C) Uracil (D) Thymine.

18. Chargaff's rule states that in DNA structure the following molar equivalences exist

(A) A = T and G = C

(B) A = G and C = T

(C) A = C and G = T

(D) A and T = deoxyribose.

19. The following statements on the DNA double helix are true except

(A) The strands run in antiparallel direction

(B) The strands are held together by hydrogen bonds

(C) The strands are composed of ribonucleotides

(D) The hydrogen bonds are formed between complementary base pairs.

20. The number of hydrogen bonds formed between guanine and cytosine

(A) 1 (B) 2 (C) 3 (D) 4.

21. Among the different types of DNA structures, the following one is most predominant under physiological conditions.

(A) A-form (B) B-form

(C) E-form (D) Z-form.

22. The following group of proteins are associated with DNA structure

(A) Albumins (B) Globulins

(C) Histones (D) Globins.

23. Orcinol colour reaction identifies one of the following components of RNA

(A) Adenine (B) Guanine

(C) Deoxyribose (D) Ribose.

24. Heterogenous RNA (hnRNA) is the precursor for the formation of

(A) Protein (B) mRNA
(C) tRNA (D) rRNA.

25. The 3′→5′ end of transfer RNA is capped with

(A) CCA (B) CAC (C) ACC (D) ACA.

Answers

1. (**B**) 2. (**C**) 3. (**B**). 4. (**D**) 5. (**B**) 6. (**A**) 7. (**C**) 8. (**D**) 9. (**A**) 10. (**B**) 11. (**C**) 12. (**A**) 13. (**C**) 14. (**A**) 15. (**A**) 16. (**D**) 17. (**D**) 18. (**A**)19. (**C**) 20. (**C**) 21. (**B**) 22. (**C**). 23. (**D**) 24. (**B**) 25. (**C**).

For clinical case studies related to Nucleotide Metabolism, refer Chapter 17.

VIVA/Short Questions

1. What are nucleic acids?

Nucleic acids (DNA and RNA) are polymers of nucleotides (polynucleotides).

2. What is DNA and what are its functions?

DNA is a polymer of deoxyribonucleotides which is organized into genes, the fundamental units of genetic information. The genes control the protein synthesis through the mediation of RNA.

3. What are the major types of RNA?

Messenger RNA, ribosomal RNA, transfer RNA.

4. What do you understand by a nucleotide?

Composed of a nitrogenous base, a pentose sugar and a phosphate. Nucleotides are the building blocks of nucleic acids (DNA and RNA).

5. What are the main functions of nucleotides?

The structural components of several coenzymes (FAD, NAD^+, $NADP^+$), energy currency of the cell (ATP). Certain second messengers of hormonal action (cyclic AMP, cyclic GMP) are nucleotides.

6. Name some biologically important nucleotides other than that found in nucleic acid structures.

FAD, NAD^+, $NADP^+$, coenzyme A, S-adenosyl-methionine, UDP-glucose.

7. What is tautomerism?

The existence of a molecule in a keto (lactam) and enol (lactim) forms.

8. Name the sugars found in nucleic acids.

The five carbon monosaccharides (pentoses) are found in nucleic acids. DNA contains deoxyribose while RNA contains ribose.

9. What is Chargaff's rule?

The molar equivalence between the purines and pyrimidines in DNA structure i.e., A = T; G = C.

10. Briefly describe Watson and Crick model of DNA structure.

Consists of two polydeoxy ribonucleotide strands running antiparallel and twisted around each other. The strands are held together by hydrogen bonds formed by complementary base pairs (A = T, C ≡ G).

11. What is heterogenous nuclear RNA (hnRNA)?

mRNA synthesized in the nucleus of eukaryotes.

6. Enzymes

MCQs

1. The presence of one of the following is absolutely essential for the chemical reactions to occur in the living cells

(A) Carbohydrates (B) Enzymes
(C) Proteins (D) Nucleic acids.

2. All of the following are intracellular enzymes except

(A) Trypsin (B) Fumarase
(C) Hexokinase
(D) Alcohol dehydrogenase.

3. As per the IUB system, enzymes are grouped into the following number of major classes

(A) 2 (B) 4 (C) 7 (D) 8.

4. The following is the correct order of enzyme classes

(A) Oxidoreductases, hydrolases, lyases, transferases, ligase, isomerases, translocases
(B) Oxidoreductases, transferases, lyases, hydrolases, isomerases, ligases, translocases
(C) Oxidoreductases, transferases, hydrolases, lyases, isomerases, ligases, translocases
(D) Oxidoreductases, hydrolases, transferases, lyases, ligases, isomerases, translocases

5. One of the following class of enzymes is mostly involved in biosynthetic reactions

(A) Lyases (B) Ligases
(C) Transferases (D) Hydrolases.

6. Besides the proteins, one of the following has been shown to act as enzymes

(A) DNA (B) Carbohydrates
(C) RNA (D) Lipids.

7. The name coenzyme refers to the following component of enzyme

(A) Non-protein, inorganic, dialysable

(B) Non-protein, organic, dialysable

(C) Inactive polypeptide chain

(D) Active protein chain.

8. One of the following is not a multienzyme complex

(A) Pyruvate dehydrogenase

(B) Fatty acid synthase

(C) Prostaglandin synthase

(D) L-amino acid oxidase.

9. The most essential pre-requisite for enzyme activity

(A) Concentration of enzyme

(B) Concentration of substrate

(C) Contact between enzyme and substrate

(D) Contact between coenzyme and substrate.

10. Michaelis-Menten constant (K_m) refers to the following in enzyme kinetics

(A) The concentration of substrate to produce maximum velocity

(B) The concentration of the substrate to produce half-maximal velocity

(C) The concentration of the enzyme to produce half-maximal velocity

(D) The concentration of the coenzyme to produce half-maximal velocity.

11. K_m value indicates the affinity between

(A) Enzyme and coenzyme

(B) Enzyme and substrate

(C) Coenzyme and substrate

(D) Enzyme and cofactor.

12. The graphic representation of Lineweaver Burk reciprocal plot is

(A) Bell shaped (B) Hyperbola

(C) Straight line (D) Sigmoidal.

13. The optimum pH for urease activity

(A) 40°C (B) 50°C (C) 60°C (D) 80°C.

14. At higher temperature, most of the mammalian enzymes become inactive due to

(A) Liberation of amino acids

(B) Loss of sulfhydryl groups

(C) Cleavage of peptide bonds

(D) Loss of tertiary structure.

15. The following cation/anion is required by amylase

(A) Mg^{2+} (B) Cl^- (C) Na^+ (D) SO_4^{2-}.

16. Active site consists of

(A) A linear sequence of amino acids

(B) Amino acids far from each other in enzyme structure

(C) All the 20 amino acids

(D) Only 10 essential amino acids.

17. One of the following amino acids is most frequently found at the active site of several enzymes

(A) Glycine (B) Leucine

(C) Serine (D) Threonine.

18. In one of the following categories of enzyme inhibition, the inhibitor closely resembles the substrate in structure

(A) Competitive (B) Non-competitive

(C) Un-competitive (D) Allosteric.

19. The competitive inhibitor is characterized by one of the following

(A) Increase in K_m (B) Increase in V_{max}

(C) Decrease in K_m (D) Decrease in V_{max}.

20. One of the following is the most potent competitive inhibitor of succinate dehydrogenase

(A) Glutaric acid (B) Oxalic acid

(C) Malonic acid (D) Succinic acid.

21. The non-competitive inhibitor is characterized by the following except

(A) The inhibitor does not bind at the active site

(B) The inhibitor binds with the enzyme as well as the enzyme-substrate complex

(C) V_{max} is lowered

(D) K_m is unchanged.

22. One of the following drugs employed in the treatment of leukemia and certain other cancers is a competitive inhibitor of the enzyme dihydrofolate reductase

(A) Allopurinol (B) Amphetamine

(C) Sulfanilamide (D) Aminopterin.

23. All of the following classes of enzymes exhibit stereospecificity except

(A) Oxidoreductases (B) Hydrolases

(C) Isomerases (D) Transferases.

24. One of the following enzymes is a good example for absolute substrate specificity

(A) Hexokinase (B) Glucokinase

(C) Chymotrypsin (D) Lipase.

25. Most of the coenzymes are derivatives of

(A) Water soluble vitamins

(B) Fat soluble vitamins (C) Proteins

(D) Nucleic acids.

26. One of the following is a non-vitamin coenzyme

(A) Coenzyme A (B) Lipoic acid

(C) S-adenosylmethionine (D) $NADP^+$.

27. The specificity of the enzyme is mostly dependent on the

(A) Apoenzyme (B) Coenzyme

(C) Prosthetic group (D) Cofactor.

28. Enzymes act by

(A) Increasing the activation energy of the reactants

(B) Decreasing the activation energy of the reactants

(C) Increasing the equilibrium constant

(D) Decreasing the activation energy of the products.

29. Fischer's template theory of enzyme-substrate complex formation is characterized by

(A) Rigid and pre-shaped active site

(B) Flexible active site

(C) Flexible substrate

(D) Rigid substrate.

30. Koshland's model of enzyme-substrate complex formation is characterized by

(A) Rigid and pre-shaped active site

(B) Substrate-inducible flexible active site

(C) Rigid structure of substrate

(D) Flexible structure of substrate.

31. K-class of allosteric enzymes are characterized by

(A) Altered K_m and V_{max}

(B) Unaltered K_m and V_{max}

(C) Altered K_m and unaltered V_{max}

(D) Unaltered K_m and altered V_{max}.

32. The following type of inhibition helps in the control of metabolic pathways

(A) Allosteric inhibition

(B) Competitive inhibition

(C) Non-competitive inhibition

(D) Un-competitive inhibition.

33. One of the following is an example of active enzyme in a dephosphorylated state

(A) Glycogen phosphorylase

(B) Glycogen synthase

(C) Succinate dehydrogenase

(D) Hexokinase.

34. The enzymes of the following metabolic pathways are localized in the cytosol except

(A) Glycolysis

(B) Hexose monophosphate shunt

(C) Fatty acid synthesis

(D) Fatty acid oxidation.

35. In general, the rate limiting enzymes of metabolic pathways have

(A) Short half-lives (B) Long half-lives

(C) Half-lives equal to that of other enzymes

(D) None of the above.

36. The group of enzymes whose concentrations/activities are regulated as per the cellular needs are called

(A) Constitutive enzymes

(B) Adaptive enzymes

(C) House-keeping enzymes

(D) Allosteric enzymes.

37. The synthesis of all of the following enzymes is increased (induction) by cortisol except

(A) Pyruvate carboxylase

(B) Tryptophan pyrrolase

(C) Hexokinase

(D) Tyrosine aminotransferase.

38. Enzymes of living system are usually expressed in terms of

(A) Concentration (μg/mg/tissues)

(B) Activities (katal or SI units)

(C) Concentration (mmol/l)

(D) None of the above.

39. It is believed that the following compounds were functioning as enzymes before the occurrence of protein enzymes

(A) DNAs (B) RNAs

(C) Carbohydrates (D) Lipids.

40. One of the following enzymes is used as a therapeutic agent for clearing the blood clots

(A) Asparaginase (B) Fibrinogen

(C) Streptokinase (D) Peroxidase.

41. One of the following is a plasma non-functional enzyme

(A) Lipoprotein lipase (B) Choline esterase
(C) Ceruloplasmin
(D) Lactate dehydrogenase.

42. Serum alkaline phosphatase is elevated in

(A) Obstructive jaundice
(B) Hepatitis (C) Pancreatitis
(D) Myocardial infarction.

43. The following enzyme is increased in the serum in chronic alcoholic patients

(A) Aldolase (B) Alcohol dehydrogenase
(C) Lactate dehydrogenase
(D) γ-glutamyl transpeptidase.

44. Isoenzymes are characterized by

(A) Similarity in chemical structure
(B) Identical immunological properties
(C) Catalysing the same reaction
(D) Possessing same K_m and V_{max} values.

45. The existence of alkaline phosphatase in the form of isoenzyme is attributed to the

(A) Presence of different subunits
(B) Synthesis by different genes
(C) Difference in carbohydrate content
(D) None of the above.

46. Lactate dehydrogenase (LDH) exists in the following number of isoenzymes

(A) 2 (B) 3 (C) 4 (D) 5.

47. One of the following isoenzymes of LDH is inhibited by pyruvate in the heart muscle

(A) LDH_1 (B) LDH_2 (C) LDH_3 (D) LDH_5.

48. All of the following serum enzymes are elevated in myocardial infarction except

(A) Creatine phosphokinase
(B) Aspartate transaminase
(C) Alkaline phosphatase
(D) Lactate dehydrogenase.

49. The enzyme acid phosphatase is elevated in serum in

(A) Obstructive jaundice
(B) Rickets (C) Hepatitis
(D) Prostatic carcinoma.

50. The following serum enzyme(s) is (are) increased in muscular diseases

(A) Aldolase
(B) Creatine phosphokinase
(C) Aspartate transaminase
(D) All of the above.

51. The physician prescribed the disulfiram drug for the treatment of alcoholism. Disulfiram specifically inhibits the following enzyme

(A) Alcohol dehydrogene
(B) Aldehyde dehydrogenase
(C) Pruvate dehydrogenase
(D) Lactate dehydrogenase.

52. The physician administers the following enzyme for clearing blood clots

(A) Asparaginase (B) Streptokinase
(C) Lipase (D) α-Antitrypsin.

Answers

1. **(B)** 2. **(A)** 3. **(C)** 4. **(C)** 5. **(B)** 6. **(C)** 7. **(B)** 8. **(D)** 9. **(C)** 10. **(B)** 11. **(B)** 12. **(C)** 13. **(C)** 14. **(D)** 15. **(B)** 16. **(B)** 17. **(C)** 18. **(A)** 19. **(A)** 20. **(C)** 21. **(A)** 22. **(D)** 23. **(C)** 24. **(B)** 25. **(A)** 26. **(C)** 27. **(A)** 28. **(B)** 29. **(A)** 30. **(B)** 31. **(C)** 32. **(A)** 33. **(B)** 34. **(D)** 35. **(A)** 36. **(B)** 37. **(C)** 38. **(B)** 39. **(B)** 40. **(C)** 41. **(D)** 42. **(A)** 43. **(D)** 44. **(C)** 45. **(C)** 46. **(D)** 47. **(A)** 48. **(C)** 49. **(D)** 50. **(D)** 51. **(B)** 52. **(B)**.

Clinical Case Study 6.1
(Source : Biochemistry, Page 115)

A 55-year old man was brought to the hospital with severe chest pain, breathlessness and vomiting. He could be rushed to the city hospital 5 hours after the onset of chest pain. His laboratory data

Parameter (serum)	*Subject*	*Reference range*
Creatine phosphokinase (CPK)	600 U/l	50-200 U/l
Aspartate transaminase (AST)	85 U/l	10-50 U/l
Lactate dehydrogenase (LDH)	315 U/l	50-200 U/l

Further, the isoenzymes CPK_2(MB) and LDH_1 were highly elevated.

1. **What is the probable diagnosis?**
2. **Which enzyme is the earliest diagnostic marker of this disease?**
3. **Name the isoenzyme of LDH elevated in this disorder?**

Ans. :

1. **Myocardial infarction.**
2. CPK starts to rise at 4-6 hours and reaches a peak value within 24-30 hours. Thus, CPK is the earliest marker enzyme for the diagnosis of myocardial infarction.
3. Isoenzyme LDH_1.

Clinical Case Study 6.2
(Source : Biochemistry, Page 115)

A 60-year old man had mild chest pain and breathlessness. He ignored it. His son came to know about this and could bring him to the city hospital only on the 3rd day after he had the symptoms. The man's laboratory findings are given hereunder.

Parameter (serum)	*Subject*	*Reference range*
Creatine phosphokinase (CPK)	120 U/l	50-200 U/l
Aspartate transaminase (AST)	350 U/l	10-50 U/l
Lactate dehydrogenase (LDH)	620 U/l	200-450 U/l
LDH electrophoresis	LDH_1 elevated	

1. **What is the probable diagnosis?**
2. **Which enzymes support the diagnosis of this disease?**
3. **Why CPK is almost normal in this disease?**

Ans. :

1. **Myocardial infarction.**
2. Elevation in the serum LDH (particularly LDH_1) and AST support the diagnosis.
3. CPK is almost normal since the patient was brought to the hospital on the third day after he developed complications (Note : CPK rises at 4-6 hours and reaches peak value within 24-30 hours).

Clinical Case Study 6.3
(Source : Biochemistry, Page 110)

A 50-year old man, an occasional alcoholic, had the complaint of severe abdominal pain following a large meal or alcohol intake. He also had the symptoms of nausea and vomiting. His laboratory data are given hereunder.

Parameter	*Subject*	*Reference range*
Serum amylase	650 U/l	30-150 U/l
Serum lipase	450 U/l	30-150 U/l
Urine amylase	1350 U/l	200-800 U/l

1. **What is the probable diagnosis?**
2. **How is amylase activity linked with this disease?**
3. **What are probable causes of this disease?**

Ans. :

1. **Pancreatitis**
2. Serum amylase activity is elevated on the first day of the disease and falls rapidly due to renal clearance.
3. Refer chapter 8, page 180.

Clinical Case Study 6.4
(Source : Biochemistry, Page 112)

A 70-year old man had the complaints of frequent urination and pain in the lower abdomen. He had a sensation of obstruction in his urine flow. His rectal examination revealed the presence of hard mass. The laboratory findings of the man

Parameter	*Subject*	*Reference range*
Serum acid phosphatase	30 U/l	0.5-5 U/l
Serum alkaline phosphatase	60 U/l	50-100 U/l
Serum prostatic specific antigen (PSA)	15 ng/ml	< 4 ng/ml

1. **What is the probable diagnosis?**
2. **Which is more reliable indicator of this disease?**

Ans. :

1. **Cancer of prostate gland.**
2. Estimation of serum PSA (elevated) is highly specific for prostate carcinoma.

For more clinical case studies, refer 7.2, 9.1, 14.5, 17.1

VIVA/Short Questions

1. **What are enzymes?**

Biocatalysts synthesized by living cells.

2. **Name the enzyme that was first isolated, crystallized and identified as a protein.**

Urease.

3. **How are enzymes classified?**

1. Oxidoreductases 2. Transferases 3. Hydrolases 4. Lyases 5. Isomerases 6. Ligases 7. Translocases.

4. **What is a holoenzyme?**

The functional unit of an enzyme made up of an apoenzyme (the protein part) and a coenzyme (the non-protein part).

5. **What do you understand by a prosthetic group?**

The non-protein moiety which tightly (covalently) binds to the apoenzyme.

6. **Name some important factors affecting enzyme activity.**

Concentration of enzyme, concentration of substrate, temperature, pH, presence of products, activators and inhibitors.

7. How does substrate concentration influence enzyme activity?

As the substrate concentration increases, there is a gradual increase in the velocity of enzyme reaction which then platues.

8. What is K_m?

K_m or Michaelis-Menten constant refers to the substrate concentration (mole/l) to produce half-maximal velocity in an enzyme catalyzed reaction.

9. How is K_m value related to enzyme affinity?

A low K_m value indicates a strong affinity between enzyme and substrate, whereas a high K_m value reflects a weak affinity.

10. What is a Lineweaver-Burk double reciprocal plot?

A straight line graphic representation depicting the relation between substrate concentration and enzyme velocity. This plot is employed for the calculation of K_m values for enzymes.

11. Give some examples of enzymes requiring inorganic metals for their activity.

ATPase (Mg^{2+} and Ca^{2+}); enolase (Mg^{2+}); xanthine oxidase (Mo). amylase (Cl^-).

12. What is active site?

The small region at which the substrate(s) binds with the enzyme.

13. What are the characteristic features of active site?

It is flexible in nature and it exists due to the tertiary structure of protein. Acidic, basic and hydroxyl amino acids are found at the active site.

14. What do you understand by enzyme inhibitor?

A substance which binds with the enzyme and brings about a decrease in catalytic activity.

15. What are the different types of enzyme inhibition?

(a) Reversible inhibition (b) Irreversible inhibition (c) Allosteric inhibition.

16. What are the types of enzyme specificities?

(a) Stereospecificity (b) Reaction specificity (c) Substrate specificity

17. What do you understand by a coenzyme?

The non-protein, organic, low molecular weight and dialysable substance associated with enzyme function.

18. Name some non-vitamin coenzymes.

ATP, CDP, UDP.

19. What do you understand by nucleotide coenzymes.

Some of the coenzymes possess a nitrogenous base, a sugar and phosphate. e.g., NAD^+, $NADP^+$, FMN, FAD, coenzyme A, UDPG.

20. What are the important factors that regulate enzyme activities in the living system?

Allosteric regulation, activation of latent enzymes, compartmentation of metabolic pathways, control of enzyme synthesis, enzyme degradation, occurrence of isoenzymes.

21. What do you understand by feedback regulation?

The process of inhibiting the first step by the final product (end product) in a series of enzyme-catalyzed reactions of a metabolic pathway.

22. What are proenzymes?

Some enzymes are synthesized in an inactive form, referred to as proenzymes (or zymogens).

23. What do you understand by constitutive enzymes?

The group of enzymes whose synthesis is not controlled and their concentrations are fairly constant e.g., citric acid cycle enzymes.

24. What is induction and what is repression?

The term induction is used to represent increased synthesis of enzyme (protein) while repression indicates its decreased synthesis.

25. How are enzymes expressed with regard to their catalytic function?

In terms of their activity with regard to their catalysis e.g., Katal, International Units.

26. What is a Katal?

A Katal (kat) represents the conversion of one mole of the substrate per second (mol/sec).

27. What do you understand by SI Unit in enzymes?

One System International (SI) Unit or one International Unit (IU) refers to the activity of the enzymes that catalyses the conversion of one micromol of the substrate per minute.

28. What are ribozymes?

The RNAs that can function as enzymes non-protein enzymes.

29. Give some examples of enzymes used as therapeutic agents.

Streptokinase is used for clearing blood clots.
Asparaginase in the treatment of leukemias.

30. What do you understand by plasma functional enzymes?

Certain enzymes normally present in the plasma, perform specific functions e.g., lipoprotein lipase, plasmin, choline esterase.

31. What are plasma non-functional enzymes and what is their importance?

These enzymes are either totally absent or present at a low concentration in plasma compared to their levels found in tissues. Estimation of plasma non-functional enzymes is important for the diagnosis and prognosis of several diseases.

32. Name some important serum enzymes useful for the diagnosis of diseases.

Amylase	Acute pancreatitis
Serum glutamate pyruvate transaminase (SGPT)	Liver diseases (hepatitis)
Serum glutamate oxaloacetate transaminase (SGOT)	Heart attacks (myocardial infarction)
Alkaline phosphatase	Rickets, obstructive jaundice
Acid phosphatase	Cancer of prostate gland
Lactate dehydrogenase (LDH)	Heart attacks, liver diseases
Creatine phosphokinase (CPK)	Myocardial infarction (early marker)
Aldolase	Muscular dystrophy
5′-Nucleotidase	Hepatitis
γ-Glutamyl transpeptidase (GGT)	Alcoholism

33. What are isoenzymes (isozymes)?

The multiple forms of an enzyme catalysing the same reaction.

34. What do you know about the isoenzymes of lactate dehydrogenase?

Lactate dehydrogenase (LDH) has five distinct isoenzymes ($LDH_1 ... LDH_5$). Each one is an oligomeric protein composed of 4 subunits (M and/or H). Serum LDH_1 is elevated in myocardial infarction while LDH_5 is increased in liver diseases.

35. What is the importance of creatine phosphokinase isoenzymes?

Creatine kinase (CK) or creatine phosphokinase (CPK) exists as 3 isoenzymes. Elevation of CPK_2 (MB) in serum is an early reliable diagnostic indication of myocardial infarction.

36. Name some serum enzymes commonly estimated in the laboratory for the diagnosis of myocardial infarction.

Creatine phosphokinase (early marker), aspartate transaminase, lactate dehydrogenase.

37. Which are serum enzymes increased in liver diseases?

Alanine transaminase, aspartate transaminase, lactate dehydrogenase, alkaline phosphatase.

7. Vitamins

MCQs

1. During the course of evolution, the higher animals have almost lost the ability to synthesize one of the following group of compounds

(A) Carbohydrates (B) Lipids
(C) Vitamins (D) Nucleic acids.

2. The following two vitamins are regarded as water soluble hematopoietic

(A) E and K (B) Folic acid and B_{12}
(C) B_1 and B_6 (D) B_2 and biotin.

3. In humans, it is believed that the intestinal bacterial synthesis and their absorption of the following two vitamins adequately meets the body requirements

(A) B_1 and B_6 (B) K and E
(C) D and biotin (D) Biotin and K.

4. All the fat soluble vitamins are composed of a common carbon unit namely

(A) Isoprene (B) Isoalloxazine
(C) β-ionone (D) Tocol.

5. The following fat soluble vitamins do not perform coenzyme function except

(A) A (B) D (C) E (D) K.

6. Generally, water soluble vitamins are not stored in the body except

(A) B_1 (B) B_6 (C) B_{12} (D) Biotin.

7. The ring structure present in vitamin A is referred to as

(A) Isoprene (B) β-ionone
(C) Cyclopentanoperhydrophenanthrene
(D) Pyridine.

8. The number of isoprenoid units present in retinol

(A) 1 (B) 2 (C) 3 (D) 4.

9. One of the following elements plays an important role in the mobilization of retinol from liver

(A) Calcium (B) Potassium
(C) Zinc (D) Molybdenum.

10. One of the following vitamin A participates in Wald's visual cycle

(A) Retinol (B) Retinal

(C) Retinoic acid (D) β-carotene.

11. In certain animals for which night vision is more important, the retina of the eye possesses

(A) Equal number of rods and cones

(B) More rods and less cones

(C) More cones and less rods

(D) Only cones.

12. In rhodopsin, vitamin A is in the form of

(A) All-*trans*-retinal (B) 11-*cis*-retinal

(C) 11-*cis*-retinol (D) All-*trans*-retinol.

13. One of the following proteins is involved in the transfer of responses (in visual cycle) from the retina to the visual cortex of the brain

(A) Retinol binding protein (B) Transducin

(C) Iodopsin D) Cyanopsin.

14. The discovery of coenzyme Q was originally made in animals in the following vitamin deficiency

(A) Vitamin K (B) Thiamine

(C) Vitamin E (D) Vitamin A.

15. Increased consumption of β-carotene is associated with decreased incidence of heart attacks and certain cancers. In these cases β-carotene acts as a/an

(A) Precursor for vitamin A

(B) Antioxidant (C) Hormone

(D) Regulator of protein biosynthesis.

16. Vitamin D in its structure resembles one of the following compounds

(A) Retinol (B) Vitamin K

(C) Sterol (D) Tocopherol.

17. One of the following compounds, on exposure to sunlight (UV rays), is converted to cholecalciferol in the skin

(A) Ergosterol (B) Cholesterol

(C) Lanosterol (D) 7-dehydrocholesterol.

18. Biologically, the most active form of vitamin D

(A) Cholecalciferol

(B) 1,24-dihydroxycholecalciferol

(C) 1,25-dihydroxycholecalciferol

(D) 24,25-dihydroxycholecalciferol.

19. The number of hydroxyl groups present in 1, 25-dihydroxycholecalciferol

(A) 1 (B) 2 (C) 3 (D) 4.

20. The concentration of plasma 1,25-dihydroxycholecalciferol is regulated by circulating levels of

(A) Calcium (B) Phosphate

(C) Parathyroid hormone

(D) All of the above.

21. Calcitriol regulates the plasma levels of calcium and phosphate by acting on

(A) Intestine, skeletal muscle and bone

(B) Intestine, kidney and bone

(C) Intestine, bone and heart

(D) Kidney, bone and skeletal muscle.

22. Calcitriol increases the concentration of calcium binding protein in the intestinal cells. Calcitriol acts as a/an

(A) Phosphorylating agent

(B) Polypeptide hormone

(C) Steroid hormone (D) Antioxidant.

23. Vitamin D is a considered as a hormone bacause

(A) Cholecalciferol is synthesized in the skin

(B) Calcitriol is formed in kidney

(C) Calcitriol acts as a steroid hormone

(D) All of the above.

24. One of the following is a poor source of vitamin D

(A) Egg yolk (B) Fatty fish

(C) Yeast (D) Milk.

25. Vitamin D deficiency in children causes rickets which is characterized by

(A) Demineralization of bones with reduced serum alkaline phosphatase

(B) Mineralization of the bones with increased alkaline phosphatase

(C) Demineralization of bones with increased serum alkaline phosphatase

(D) None of the above.

26. Renal rickets is characterized by

(A) Decreased synthesis of cholecalciferol

(B) Decreased synthesis of calcitriol

(C) Decreased production of 25-hydroxycholecalciferol

(D) Increased synthesis of 24,25-dihydroxycholecalciferol.

27. Hypervitaminosis of one of the following vitamins is most toxic

(A) Vitamin A (B) Vitamin D

(C) Vitamin E (D) Vitamin K.

28. Some authors regard one of the following vitamins as 'a vitamin in search of a disease'

(A) Vitamin E (B) Folic acid

(C) Vitamin B_{12} (D) Vitamin K.

29. Tocopherols are the derivatives of

(A) Isoprenoid ring (B) Tocol

(C) Sterol (D) Ribitol.

30. The biochemical functions of vitamin E are directly or indirectly related to its

(A) Oxidant properties

(B) Antioxidant properties

(C) Hydroxylation (D) Phosphorylation.

31. High consumption of one of the following vitamins protects against the development of heart diseases

(A) Vitamin A (B) Vitamin D

(C) Vitamin E (D) Biotin.

32. Vitamin E protects the development of heart diseases by

(A) Preventing the oxidation of LDL

(B) Promoting the oxidation of LDL

(C) Preventing the oxidation of HDL

(D) Preventing the oxidation of VLDL.

33. Dietary intake of vitamin E is directly related to the consumption of

(A) Saturated fatty acids

(B) Polyunsaturated fatty acids

(C) Cholesterol

(D) Triacylglycerols.

34. One of the following is the least toxic among the fat soluble vitamins

(A) Vitamin A (B) Vitamin B

(C) Vitamin E (D) Vitamin K.

35. Vitamin K is a naphthoquinone derivative and it exists in 3 different forms. Isoprenoid side chain is absent in

(A) Vitamin K_1 (phylloquinone)

(B) Vitamin K_2 (menaquinone)

(C) Vitamin K_3 (menadione)

(D) All of the above.

36. Vitamin K acts as a coenzyme and brings about a post-translational modification namely

(A) Hydroxylation (B) Carboxylation

(C) Phosphorylation (D) Methylation.

37. Vitamin K participates in the conversion of the following inactive clotting factors to active ones in the liver

(A) I, II, IX, X (B) II, V, VII, X

(C) II, VII, IX, X (D) II, VI, VII, X.

38. One of the following amino acids in certain clotting factors gets modified through the coenzyme function of vitamin K

(A) Aspartate (B) Glutamate

(C) Glutamine (D) Tyrosine.

39. Ascorbic acid (vitamin C) resembles one of the following compounds in structure

(A) Monosaccharides (B) Nucleotides

(C) Sterols (D) Amino acids.

40. Vitamin C can be synthesized from glucose via uronic acid pathway. Man and guinea pig cannot synthesize this vitamin due to the deficiency of the enzyme

(A) Glucuronidase

(B) Xylitol dehydrogenase

(C) L-gulonolactone oxidase

(D) UDP-glucose dehydrogenase.

41. Vitamin C acts as a coenzyme and participates in the post-translational modifications of the following two amino acids in collagen

(A) Glutamate, proline (B) Lysine, proline

(C) Lysine, histidine (D) Proline, histidine.

42. Vitamin C participates in the metabolism of tryptophan and tyrosine, and the synthesis of corticosteroid hormones. In all these cases, vitamin C facilitates the following type of reaction

(A) Phosphorylation (B) Methylation

(C) Hydroxylation (D) Carboxylation.

43. The following vitamins/provitamins prevent several diseases (like cancer, heart attacks) by acting as antioxidants except

(A) Vitamin A (B) β-carotene

(C) Vitamin E (D) Vitamin C.

44. Ascorbic acid deficiency results in scurvy which is characterized by

(A) Sore gums and loose teeth

(B) Anemia and fragile blood vessels

(C) Swolen joints and delayed wound healing

(D) All of the above.

45. Ascorbic acid in its metabolism forms one of the following excretory compounds which has been implicated in the formation of urinary stones
(A) Dehydroascorbate (B) Gulonate
(C) Oxalate (D) Uric acid.

46. Thiamine contains the following ring structures
(A) Pyridine and thiazole
(B) Pyrimidine and thiazole
(C) Pyrimidine and imidazole
(D) Pyrimidine and pyridine.

47. Thiamine acts through its coenzyme thiamine pyrophosphate (TPP). TPP participates in the following enzymes of carbohydrate metabolism except
(A) Pyruvate dehydrogenase
(B) α-ketoglutarate dehydrogenase
(C) Transketolase
(D) Branched chain α-keto acid dehydrogenase

48. The daily requirement of thiamine depends on the consumption of
(A) Carbohydrate (B) Fat
(C) Amino acids (D) Cholesterol.

49. One of the following enzymes of RBC can be used for the diagnosis of thiamine deficiency
(A) Transketolase (B) Transaldolase
(C) Pyruvate dehydrogenase
(D) α-ketoglutarate dehydrogenase.

50. Riboflavin contains the following ring
(A) Ribitol (B) Purine
(C) Pyrimidine
(D) 6-7 dimethyl isoalloxazine.

51. Flavin coenzymes namely FAD and FMN participate in the following type of reaction
(A) Carboxylation (B) Oxidation-reduction
(C) Hydroxylation (D) Transamination.

52. All of the following enzymes are dependent on FAD except
(A) Pyruvate dehydrogenase
(B) Succinate dehydrogenase
(C) D-amino acid oxidase
(D) L-amino acid oxidase.

53. One of the following erythrocyte enzymes can be used to assess riboflavin deficiency
(A) Glutathione reductase
(B) Pyruvate dehydrogenase
(C) Transketolase
(D) α-ketoglutarate dehydrogenase.

54. The term pellagra (literally means rough skin) has originated from one of the following languages
(A) Greek (B) Latin (C) French (D) Italian.

55. Chemically, niacin is a derivative of one of the following
(A) Pyridine (B) Pyrimidine
(C) Tocol (D) Thiazole.

56. Tryptophan is the precursor for the synthesis of NAD^+ and $NADP^+$. This pathway occurs through the formation of an intermediate
(A) Niacin (B) Quinolinate
(C) Picolinate (D) Nicotinamide.

57. NAD^+ and $NADP^+$ are the coenzymes primarily for the following class of enzymes
(A) Oxidoreductases (B) Transferases
(C) Hydrolases (D) Ligases.

58. The normal conversion ratio of tryptophan to niacin for the synthesis of niacin coenzymes
(A) 30 : 1 (B) 60 : 1(C) 90 : 1 (D) 120 : 1.

59. One of the following amino acids has been shown to inhibit the synthesis of NAD^+ and $NADP^+$ from tryptophan
(A) Valine (B) Isoleucine
(C) Leucine (D) Tyrosine.

60. Niacin is administered in high doses for the following therapeutic purpose(s)
(A) To inhibit lipolysis in adipose tissue
(B) To reduce the synthesis of triacylglycerols in liver
(C) To decrease the serum cholesterol level
(D) All of the above.

61. Vitamin B_6 compounds are the derivatives of
(A) Pyridine (B) Pyrimidine
(C) Imidazole (D) None of the above.

62. Pyridoxal phosphate, the coenzyme of vitamin B_6, participates in the following biochemical reactions
(A) Transamination (B) Decarboxylation
(C) Transsulfuration (D) All of the above.

63. One of the following vitamins participates in the synthesis of biogenic amines (serotonin, GABA, epinephrine, and norepinephrine)
(A) Thiamine (B) Riboflavin
(C) Pyridoxine (D) Cyanocobalamin.

64. In vitamin B_6 deficiency, the following metabolite is excreted in urine
(A) Pyruvic acid (B) Xanthurenic acid
(C) Kynurenine (D) Methylmalonic acid.

65. The enzyme glycogen phosphorylase contains pyridoxal phosphate where the latter performs the following function
(A) Participates in phosphorylation reaction
(B) Brings about transamination
(C) Stabilizes the enzyme structure
(D) None of these.

66. The neurological manifestations observed in vitamin B_6 deficiency are due to an impairment in the following function of PLP
(A) Transamination (B) Decarboxylation
(C) Deamination (D) Transsulfuration.

67. One of the following drugs, used in the treatment of tuberculosis, causes vitamin B_6 deficiency
(A) Pencillamine (B) Rifampicin
(C) Isoniazid (D) Streptomycin.

68. Chemically, biotin is formed by the fusion of
(A) Pyridine and thiophene
(B) Imidazole and thiophene
(C) Imidazole and pyridine
(D) Imidazole and pyrimidine.

69. Biotin participates in the following category of reactions
(A) Carboxylation (B) Decarboxylation
(C) Deamination
(D) None of the above.

70. One of the following enzymes of gluconeogenesis is dependent on biotin
(A) Pyruvate kinase
(B) Glucose 6-phosphatase
(C) Pyruvate carboxylase
(D) Malate dehydrogenase.

71. Biotin participates in CO_2 incorporation reactions in metabolism. One of the following however, is an exception
(A) Synthesis of carbamoyl phosphate
(B) Formation of malonyl CoA
(C) Synthesis of methylmalonyl CoA
(D) Formation of methylglutaconyl CoA.

72. The raw egg white contains the following protein which binds to biotin and blocks its absorption from the intestine
(A) Biocytin (B) Avidin
(C) Transferrin (D) Ceruloplasmin.

73. One of the following amino acids is a component of panthotenic acid (coenzyme A)
(A) Glycine (B) α-alanine
(C) β-alanine (D) Proline.

74. All of following enzymes require coenzyme A except
(A) Pyruvate dehydrogenase
(B) Lactate dehydrogenase
(C) α-ketoglutarate dehydrogenase
(D) Thiokinase.

75. The coenzyme regarded as the coenzyme of metabolic integration is
(A) Coenzyme A
(B) Thiamine pyrophosphate
(C) Tetrahydrofolate
(D) Pyridoxal phosphate.

76. Pantothenic acid (not coenzyme A) is a component of one of following complex enzymes
(A) Pyruvate dehydrogenase
(B) Fatty acid synthase
(C) Glycogen phosphorylase
(D) Lactate dehydrogenase.

77. Chemically, folic acid is composed of
(A) Pyridine, p-amino benzoic acid, glutamate
(B) Pyrimidine, p-amino benzoic acid, glutamate
(C) Pteridine, p-amino benzoic acid, glutamate
(D) Imidazole, p-aminobenzoic acid, aspartate.

78. In the circulation, one of the following forms of tetrahydrofolate is most abundant in concentration
(A) N^5-methyltetrahydrofolate
(B) N^{10}-methyltetrahydrofolate
(C) N^5, N^{10}-methylenetetrahydrofolate
(D) N^5-formiminotetrahydrofolate.

79. The following compounds are synthesized in one carbon metabolism through the participation of tetrahydrofolate
(A) Purine nucleotides (in turn incorporated into DNA and RNA)
(B) Pyrimidine nucleotide (dTMP) involved in the synthesis of DNA
(C) Glycine, serine, ethanolamine and choline
(D) All of the above.

80. One of the following is a poor source of folic acid
(A) Milk (B) Leafy vegetables
(C) Liver (D) Cereals.

81. Formiminoglutamate (FIGLU) accumulates in the body and gets excreted into urine in folic acid deficiency. FIGLU is produced in the metabolism of

(A) Glutamate (B) Aspartate
(C) Histidine (D) Arginine.

82. The folic acid antagonists namely aminopterin, amethopterine, and methotrexate inhibit the enzyme dihydrofolate reductase. This is a classical example of

(A) Competitive inhibition
(B) Allsoteric inhibition
(C) Non-competitive inhibition
(D) Uncompetitive inhibition.

83. Vitamin B_{12} is synthesized by

(A) Animals only (B) Plants only
(C) Animals and plants
(D) Microorganisms only.

84. The following statement (s) is (are) true about the structure of vitamin B_{12}

(A) It has a corrin ring with a central cobalt atom
(B) Each corrin ring is composed of 4 pyrrole units
(C) Cobalt is bound to dimethylbenzimidazole below the plane of corrin ring
(D) All of the above.

85. The term entrinsic factor of Castle refers to

(A) The dietary source of vitamin B_{12}
(B) The glycoprotein required for B_{12} absorption
(C) The protein of B_{12} transport
(D) None of the above.

86. The liver stores B_{12} mostly in the form of

(A) Methylcobalamin
(B) Deoxyadenosyl cobalamin
(C) Hydroxycobalamin (D) Cyanocobalamin.

87. The following enzyme requiring methylcobalamin as coenzyme is responsible for the conversion of homocysteine to methionine

(A) Homocysteine methyltransferase
(B) Homocysteine decarboxylase
(C) Homocysteine dehydrogenase
(D) Homocystine transaminase.

88. The coenzyme required by methylmalonyl CoA mutase

(A) Methylcobalamin
(B) Deoxyadenosylcobalamin
(C) Cyanocobalamin (D) Tetrahydrofolate.

89. The cause(s) attributed to vitamin B_{12} deficiency resulting in pernicious anemia

(A) Hereditary malabsorption of vitamin B_{12}
(B) Partial or total gastrectomy
(C) Insufficient production of intrinsic factor
(D) Any one of the above.

90. The neurological manifestation observed in pernicious anemia are attributed to the accumulation of

(A) Methylmalonyl CoA
(B) Homocysteine (C) Succinyl CoA
(D) Methylmalonic acid.

91. Estimation of the following metabolite (elevated) can be used to assess vitamin B_{12} deficiency

(A) Propioni acid (B) Xanthurenic acid
(C) Methylmalonic acid (D) Pyruvic acid.

92. The deficiency of the following two vitamins causes similar type of anemia

(A) Vitamin B_{12} and niacin
(B) Vitamin B_{12} and niacin
(C) Vitamin B_{12} and biotin
(D) Vitamin B_{12} and folic acid.

93. The enzyme that links the function of vitamin B_{12} and folic acid

(A) Homocysteine methyltransferase
(B) Cystathioninase
(C) Cysteine decarboxylase
(D) Methionine methyltransferase.

94. A 50 year old woman, a patient of tuberculosis, was given the drug isoniazed (isonicolinic acid hydrazide INH). She developed neurological manifestations (depression, nervousness, confusion etc.). Her urine contained high concentration of xanthurenic acid. The drug INH results in the deficiency of the following vitamin.

(A) Niacin (B) Vitamin B_6
(C) Vitamin B_{12} (D) Vitamin C

95. A 25 year old woman, on being tested positive for pregnancy, approaches her family physician. The following nutrient supplements are recommended by the doctor for the healthy growth of the fetus.

(A) Vitamins A and B (B) Vitamin K and Se
(C) Riboflavin and copper
(D) Iron and folate

96. Vitamin C is prescribed for quick wound healing by the doctors, since it promotes the synthesis of

(A) Keratin (B) Elastin
(C) Collagen (D) Globulins

97. A 10 year old boy on strict vegetarian diet presented with severe vomiting, muscle weakness, severe anemia and lethargy. Peripheral smear showed macrocytic anemia and increased MCV. The probable diagnosis is deficiency of

(A) Vitamin B_6 (B) Vitamin B_1
(C) Vitamin A (D) Vitamin B_{12}

98. A 3 year old boy had failure to thrive with unusual gait. He had features of bow legs, widened wrists and dental caries. Serum alkaline phosphatase is elevated, with low calcium level. What is the diagnosis?

(A) Osteomalacia (B) Scurvy
(C) Rickets (D) Hyperparathyroidism

99. A 25 year old woman presented with progressive loss of vision, manifested as night blindness. Gradually, the vision during day light also dimnished. Ophthalmologist examined and observed bitot spots. This is due to the defeciency of

(A) Vitamin C (B) Vitamin A
(C) Vitamin D (D) Vitamin K

100. A 5 year old girl presented with swollen knee joint and loss of teeth easily and had a injury for which wound was not healing. What is the diagnosis

(A) Rickets (B) Hemophilia
(C) Pellagra (D) Scurvy

Answers

1. **(C)** 2. **(B)** 3. **(D)** 4. **(A)** 5. **(D)** 6. **(C)** 7. **(B)** 8. **(B)** 9. **(C)** 10. **(B)** 11. **(B)** 12. **(B)** 13. **(B)** 14. **(D)** 15. **(B)** 16. **(C)** 17. **(D)** 18. **(C)** 19. **(C)** 20. **(D)** 21. **(B)** 22. **(C)** 23. **(D)** 24. **(D)** 25. **(D)** 26. **(B)** 27. **(B)** 28. **(A)** 29. **(B)** 30. **(B)** 31. **(C)** 32. **(A)** 33. **(B)** 34. **(C)** 35. **(C)** 36. **(B)** 37. **(C)** 38.**(B)** 39. **(A)** 40. **(C)** 41. **(B)** 42. **(C)** 43. **(A)** 44. **(D)** 45. **(C)** 46. **(B)** 47. **(D)** 48. **(A)** 49. **(A)** 50. **(D)** 51. **(B)** 52. **(D)** 53. **(A)** 54. **(D)** 55. **(A)** 56. **(B)** 57. **(A)** 58. **(B)** 59. **(C)** 60. **(D)** 61. **(B)** 62. **(D)** 63. **(C)** 64. **(B)** 65. **(C)** 66. **(B)** 67. **(C)** 68. **(B)** 69. **(A)** 70. **(C)** 71. **(A)** 72. **(B)** 73. **(C)** 74. **(B)** 75. **(A)** 76. **(B)** 77. **(C)** 78. **(A)** 79. **(D)** 80. **(A)** 81. **(C)** 82. **(A)** 83. **(D)** 84. **(D)** 85. **(A)** 86. **(B)** 87. **(A)** 88. **(B)** 89. **(D)** 90. **(A)** 91. **(C)** 92. **(D)** 93. **(A)** 94. **(B)** 95. **(D)** 96. **(C)** 97. **(D)** 98. **(C)** 99. **(B)** 100. **(D)**.

Clinical Case Study 7.1
(Source : Biochemistry, Page 126)

A 12-year old boy had difficulty to see in dim light (evenings and while he enters cine theatres). His plasma retinol level was very low.

1. **What is the probable diagnosis?**
2. **Which vitamin is linked with this disorder?**
3. **What happens in severe deficiency of this vitamin?**

Ans. :

1. **Night blindness (nyctalopia)**
2. Vitamin A deficiency. Since vitamin A is closely associated with dim light vision, its deficiency causes prolonged dark adaptation time.
3. Severe deficiency of vitamin A leads to dryness of conjunctiva and cornea (xerophthalmia) and the occurrence of triangular plaques known as Bitot's spots.

Clinical Case Study 7.2
(Source : Biochemistry, Page 131)

A 6-year old boy had bone deformities such as bow legs and pigeon chest. He had a history of delayed eruption of teeth. The boy had been on a low intake of milk. The following are the laboratory findings.

Parameter (serum)	*Subject*	*Reference range*
Calcium	8.1 mg/dl	9-11 mg/dl
Phosphate	2.5 mg/dl	3-4.5 mg/dl
Alkaline phosphatase	300 U/l	50-100 U/l
25-Hydroxyvitamin D	10 ng/ml	25-50 ng/ml

1. **What is the probable diagnosis?**
2. **What is the deficient nutrient?**
3. **What is the biochemical explanation for bone deformities?**
4. **How do you explain elevated serum alkaline phosphatase?**

Ans. :

1. Classical example of **rickets**.
2. As a result of low serum calcitriol (the biochemically active form of vitamin D), calcium and phosphate levels are not maintained in the circulation, hence mineralization is impaired, ultimately causing bone deformities such as bow legs.
3. Serum alkaline phosphatase is elevated in rickets in a vain attempt to result in bone formation.

Clinical Case Study 7.3
(Source : Biochemistry, Page 137)

A 25-year old working woman developed the symptoms of spongy and sore gums, swollen joints. She used to have frequent infections (respiratory tract infections) and observed delay in healing her wounds. Nutritionist, advised her to eat fresh fruits and vegetables, besides the supplementation of vitamin C. The situation improved in about 3 months.

1. **What is the probable diagnosis? Justify.**
2. **What are the major biochemical functions of vitamin C?**
3. **How is vitamin C involved in collagen synthesis?**

Ans. :

1. The manifestations described along with the response of the woman to vitamin C supplementation clearly point that the woman was suffering from **scurvy**.
2. Vitamin C participates in the synthesis of collagen, besides its involvement as an antioxidant. The symptoms observed in scurvy are due to impairment in these functions.
3. Vitamin C plays the role of a coenzyme in hydroxylation of proline and lysine while protocollagen is converted to collagen i.e., post-translational modifications.

Clinical Case Study 7.4
(Source : Biochemistry, Page 140)

A 40-year old alcoholic man had the symptoms of mental confusion, loss of memory, rhythmical to and fro motion of eye balls. From the analysis of the cultured fibroblasts, the man was found to have the enzyme transketolase with much reduced affinity for thiamine pyrophosphate.

1. **What is the probable diagnosis?**
2. **In which metabolic pathway the enzyme transketolase participates?**
3. **What is significance of this pathway.**

Ans. :

1. This is a case of **Wernicke-Korsakoff syndrome** due to the deficiency of thiamine.
2. Hexose monophosphate shunt.
3. It generates two important products – pentoses and NADPH-needed for biosynthetic reactions (nucleotides, and lipids) and several other functions.

Clinical Case Study 7.5
(Source : Biochemistry, Page 140)

A 55-year old woman, a patient of tuberculosis, was given the drug isoniazid (isonicotinic acid hydrazide, INH). She developed neurological manifestations such as depression, irritability, nervousness. Her urine contained increased concentration of xanthurenic acid.

1. **What is the probable diagnosis?**
2. **What is the biochemical explanation for the increased excretion of xanthurenic acid?**
3. **What is the nutritional advice for TB patients on treatment with INH?**

Ans. :

1. **Drug induced vitamin B_6 deficiency.**
2. The drug isoniazid combines with pyridoxal phosphate (PLP), the coenzyme of vitamin B_6, to form inactive hydrazone derivatives which inhibit PLP dependent enzymes. The elevated excretion of xanthurenic acid indicates a blockade in tryptophan metabolism (due to impairment in the activity of kynureninase).
3. Treatment with isoniazid should be accompanied by vitamin B_6 supplementation.

Clinical Case Study 7.6
(Source : Biochemistry, Page 155)

A 23-year old woman with 3 children had the complaints of weakness and lethargy. Her hemoglobin level was 7 g/dl (normal 12-14 g/dl). Her blood was found to contain large abnormal immature erythrocytes. This woman had a highly elevated formiminoglutamate (FIGLU), in urine.

1. **What is the vitamin deficiency?**
2. **What is the probable diagnosis?**
3. **What is the explanaton for presence of large abnormal immature RBC?**
4. **In which amino acid metabolism, FIGLU is formed?**

Ans. :

1. Elevated excretion of FIGLU in urine indicates that the woman was suffering from **folate deficiency**.
2. **Macrocytic anemia** with abnormally large erythrocytes.
3. Due to a block in DNA synthesis, the maturation of RBC is slowed down leading to the increase in their size.
4. Histidine.

Clinical Case Study 7.7
(Source : Biochemistry, Page 159)

A 25-year old strict vegetarian woman, with a very low consumption of milk products, was found to be anemic (hemoglobin 7 g/dl against normal 12-14 g/dl). She was lethargic and had the complaints of numbness and tingling of fingers and toes. Her urine had an elevated excretion of methylmalonate.

1. **What is the probable diagnosis?**
2. **What is the biochemical explanation for the elevated excretion of methylmalonate?**
3. **What are the good sources of this vitamin?**
4. **Why curd is better source than milk for this vitamin?**

Ans. :

1. **Vitamin B_{12} deficiency.**
2. Refer Chapter 7, page 158.
3. Foods of animal origin – liver, kidney, milk, fish, pork, chicken.
4. Due to the synthesis of this vitamin by *Lactobacillus* in curd.

Clinical Case Study 7.8
(Source : Biochemistry, Page 131)

A teenaged boy started taking vitamin D pills 4 per day (about 1600 IU/day), besides consuming diet rich in vitamin D. He slowly developed the symptoms of loss of appetite, nausea and incresed thirst, besides loss of weight. Biochemical investigations revealed that his serum calcium was 14 mg/dl (normal 9-11 mg/dl) and phosphate 6.5 mg/dl (normal 3-4.5 mg/dl).

1. **What is the probable diagnosis?**
2. **What are the manifestations?**
3. **Which is the most toxic vitamin in overdose?**
4. **What is the active form of this vitamin?**

Ans. :

1. **Hypervitaminosis D**
2. The characteristic clinical manifestations and biochemical profile (elevation in serum calcium and phosphate). Prolonged high intake of vitamin D may lead to calcification of blood vessels and formation of kidney stones.
3. Vitamin D is the most toxic.
4. Calcitriol (1,25-dihydroxycholecalciferol).

For more clinical case studies, refer 14.6, 18.3

VIVA/Short Questions

1. **What are vitamins?**

Organic compounds required in the diet in small amounts to perform specific biological functions for growth and health of the organism.

2. **How are vitamins classified?**

Fat soluble (A, D, E, K) and water soluble (B-complex and C). The water soluble vitamins are energy releasing (B_1, B_2, niacin, B_6, biotin, pantothenic acid) and hematopoietic (folic acid and B_{12}).

3. **What do you understand by coprophagy?**

Some of the animals (e.g., rat, deer) eat their own feces, this is known as coprophagy.

4. **What are the general characteristics of fat soluble vitamins?**

The four fat soluble vitamins (A, D, E, K) are soluble in fats, oils and fat solvents (alcohol, acetone etc.). Contain one or more of isoprene units (5 carbon units). They can be stored in liver and adipose tissue.

5. **What are the general features of water soluble vitamins?**

They are heterogenous compounds, cannot be stored in the body in large amounts (except B_{12}). They form coenzymes that participate in a variety of biochemical reactions.

6. **Name the structurally related compounds which are commonly referred to as vitamin A.**

Retinol (vitamin A alcohol), retinal (vitamin A aldehyde), retinoic acid (vitamin A acid), β-carotene (provitamin A).

7. **What are the major functions of different vitamin A compounds?**

Retinal—visual cycle;

Retinol—steroid hormone, growth and differentiation;

Retinoic acid—steroid hormone, growth and differentiation;

Retinyl phosphate—glycoprotein synthesis;

β-Carotene—antioxidant.

8. **What is Wald's visual cycle?**

Primarily concerned with the dim light vision of rods (of retina of the eye).

9. **Name some pigments concerned with colour vision.**

Porphyropsin (red), iodopsin (green) and cyanopsin (blue). Cones of retina are specialized in bright and colour vision.

10. What are the deficiency manifestations of vitamin A?

Night blindness (nyctalopia) is one of the earliest symptoms. Severe vitamin A deficiency may lead to xerophthalmia, Bitot's spots and keratomalacia.

11. What do you understand by hypervitaminosis A?

Excessive intake of vitamin A. The symptoms include dermatitis, enlargement of liver, skeletal decalcification, loss of weight, loss of hair, irritability etc.

12. Name the provitamins of vitamin D.

Ergocalciferol (plant source) and cholecalciferol (animal source).

13. Which is the biologically most active form of vitamin D and how is it formed?

1,2-Dihydroxycholecalciferol (1,25- DHCC or calcitriol). It is produced from cholecalciferol on hydroxylation at 25th position (in liver) and then at 1st position (in kidney).

14. What are the factors that influence calcitriol formation?

Plasma calcium, phosphate and parathyroid hormone. Low plasma levels of calcium or phosphate increase calcitriol production.

15. What are the biochemical functions of calcitriol (1,25-DHCC)?

It regulates the plasma levels of calcium and phosphate through its action on intestine, kidney and bone. The net effect is an elevation in plasma Ca and P levels.

16. Vitamin D is a hormone and not a vitamin-justify.

The synthesis and functions of vitamin D are comparable to steroid hormones, the synthesis of cholecalciferol and calcitriol (the active form) in the body, action of calcitriol as a steroid hormone, short half-life of calcitriol and regulation of its synthesis by feedback mechanism.

17. What are the deficiency symptoms of vitamin D?

Rickets in children and osteomalacia in adults. Rickets is characterized by increased serum alkaline phosphatase.

18. What do you understand by renal rickets?

A disorder associated with decreased synthesis of calcitriol in kidney. This is mostly observed in patients with chronic renal failure.

19. What are the manifestations of hypervitaminosis D?

Demineralization of bone, increased calcium absorption from intestine resulting in hypercalcemia. Deposition of calcium in soft tissues and formation of stones in kidney.

20. What is the chemical nature of vitamin E?

Vitamin E refers to a group of tocopherols (α, β, γ, δ etc.). Among these, α-tocopherol is the most important.

21. What are the functions of vitamin E?

The biochemical functions of vitamin E are related directly or indirectly to its antioxidant property. The functions include maintenance of membrane structure and integrity of cell, prevention of peroxidation of polyunsaturated fatty acids and maintenance of germinal epithelium of gonads for proper reproductive function.

22. What is the chemical nature of vitamin K?

Vitamin K exists in 3 different forms — K_1 (phylloquinone), K_2 (menaquinone) and K_3 (menadione). All the three are naphthoquinone derivatives.

23. What are the biochemical functions of vitamin K?

Acts as a coenzyme for the carboxylation of glutamic acid residues present in proteins. It actually brings about post-translational modifications of blood clotting factors namely II, VII, IX and X.

24. Name the antagonists of vitamin K.

Dicumarol, bishydroxycoumarin, heparin, salicylates.

25. What do you know about the chemistry of vitamin C?

Vitamin C (ascorbic acid) is a hexose derivative and closely resembles monosaccharides in structure.

26. How is vitamin C synthesized in some animals? Why humans cannot synthesize?

From glucose via uronic acid pathway. However, man, cannot synthesize ascorbic acid due to the deficiency of a single enzyme namely L-gulonolactone oxidase.

27. What are the important biochemical functions of vitamin C?

Collagen synthesis, bone formation, immunological function, iron absorption.

28. How does vitamin C participate in collagen synthesis?

Acts as a coenzyme and brings about post-translational modifications while protocollagen is converted to collagen. Ascorbic acid is required for the conversion of proline and lysine (of protocollagen) respectively to hydroxyproline and hydroxylysine.

29. What are the deficiency symptoms of vitamin C?

Scurvy is characterized by spongy and sore gums, loose teeth, anemia, swollen joints, fragile blood vessels, delayed wound healing etc.

30. What do you know about the chemistry of thiamine (vitamin B_1)?

Contains a pyrimidine ring and a thiazole ring held by methylene bridges.

31. What are important biochemical functions of thiamine (coenzyme TPP)?

Intimately connected with energy releasing reactions of carbohydrate metabolism. The enzymes requiring TPP as the coenzyme are pyruvate dehydrogenase, α-ketoglutarate dehydrogenase, transketolase and branched chain α-keto acid dehydrogenase. In addition, TPP plays an important role in the transmission of nerve impulse.

32. What are the deficiency manifestations of thiamine?

Beri-beri. The early symptoms of thiamine deficiency are — loss of appetite, weakness, nausea, mental depression, peripheral neuropathy etc. Wet beri-beri, dry beri-beri, infantile beri-beri are known.

33. Name the antagonists of thiamine.

Pyrithiamine, oxythiamine.

34. What do you know about the chemistry of riboflavin?

Contains 6,7-dimethyl isoalloxazine attached to D-ribitol by a nitrogen atom.

35. What are the coenzymes of riboflavin?

Flavin adenine mononucleotide (FMN) and flavin adenine dinucleotide (FAD).

36. What are the biochemical functions of riboflavin?

FMN and FAD, participate in redox reactions responsible for energy generation.

37. What are the deficiency symptoms of riboflavin?

Cheilosis (fissures at the corners of mouth), glossitis (purplish tongue) and dermatitis.

38. What do you know about the chemistry of niacin?

Niacin is a pyridine derivative. Structurally, it is pyridine 3-carboxylic acid. The amide form of niacin is known as niacinamide or nicotinamide.

39. What are the coenzymes of niacin?

Nicotinamide adenine dinucleotide (NAD^+) and its phosphate ($NADP^+$).

40. What is the role of tryptophan in the synthesis of NAD^+ and $NADP^+$?

Contributes to the formation of NAD^+ and $NADP^+$. For the niacin coenzyme (NAD^+, $NADP^+$) synthesis, 60 milligram of tryptophan is equivalent to 1 mg of niacin (conversion ratio 60 : 1).

41. What are the biochemical functions of niacin (NAD^+ and $NADP^+$)?

The niacin coenzymes NAD^+ and $NADP^+$ are involved in a variety of oxidation-reduction reactions.

NAD^+ dependent : Lactate dehydrogenase, pyruvate dehydrogenase.

$NADP^+$ dependent : Glucose 6-phosphate dehydrogenase, malic enzyme.

42. What are the deficiency manifestations of niacin?

Pellagra (Italian : rough skin) symptoms referred to as three D's. Dermatitis, diarrhoea, dementia, and it not treated, may rarely lead to death (4th D).

43. What do you know about the therapeutic uses of niacin?

Administration of niacin in pharmacological doses (2-4 g/d; 200 times RDA) results in a number of biochemical effects in the body, not related to its function as a vitamin. These include inhibition of lipolysis in the adipose tissue, reduced synthesis of triacylglycerols in liver and decrease in LDL, VLDL and total cholesterol in serum.

44. What is the chemistry of vitamin B_6?

Vitamin B_6 collectively represents three compounds namely pyridoxine, pyridoxal and pyridoxamine. All of them are pyridine derivatives.

45. What do you know about the biochemical functions of vitamin B_6?

The active form of vitamin B_6, pyridoxal phosphate (PLP), is closely associated with the metabolism of amino acids – transamination, decarboxylation, deamination, transsulfuration.

46. What are the deficiency symptoms of vitamin B_6?

Neurological symptoms such as depression, irritability, nervousness and mental confusion, related to the decreased synthesis of biogenic amines (serotonin, GABA, norepinephrine and epinephrine).

47. What do you know about the drug induced B_6 deficiency?

Isoniazid is a drug frequently used for the treatment of tuberculosis. It combines with PLP to inhibit PLP dependent enzymes. Thus, administration of isoniazid results in B_6 deficiency.

48. What is the chemistry of biotin?

The structure is formed by fusion of imidazole and thiophene rings with valeric acid side chain.

49. What are the important biochemical functions of biotin?

Biotin serves as a carrier of CO_2 in carboxylation reactions. Thus, biotin participates in the synthesis of glucose (gluconeogenesis) and fatty acids.

50. Name the antagonists of biotin.

Desthiobiotin and biotin sulfonic acid.

51. What do you know about the chemistry of pantothenic acid and coenzyme A?

Pantothenic acid consists of two compounds, pantoic acid and β-alanine held together by a peptide linkage. Coenzyme A consists of pantothenic acid joined to β-mercaptoethanolamine at one end and adenylic acid at the other end.

52. What are the biochemical functions of pantothenic acid (coenzyme A)?

Coenzyme A is the central molecule involved in all the metabolisms (carbohydrate, lipid and amino acids). It is appropriately regarded as a coenzyme of metabolic integration.

53. What do you know about the chemistry of folic acid?

Consists of three components-pteridine ring, p-amino benzoic acid and glutamic acid (1-7 residues). Folic acid mostly has one glutamic acid residue and is known as pteroyl-glutamic acid (PGA).

54. What is the active form of folic acid and how it is formed?

Tetrahydrofolate (THF or FH_4). THF is synthesized from folic acid by the enzyme dihydrofolate reductase. The reducing equivalents are supplied by NADPH.

55. What are the biochemical functions of folic acid (tetrahydrofolate)?

Involved in the one carbon metabolism. THF serves as an acceptor or donor of one carbon units (formyl, methyl etc.) in a variety of reactions involving amino acid and nucleotide metabolism.

56. What do you understand by citrovorum factor?

N^5-formyl tetrahydrofolate i.e., formyl group attached to 5th position of THF.

57. What are the deficiency symptoms of folic acid?

Macrocytic anemia (abnormally large RBC). At the biochemical level, decreased production of purines and dTMP are observed which in turn impairs DNA synthesis.

58. What is the chemistry of vitamin B_{12}?

Vitamin B_{12} primarily consists of a corrin ring (with 4 pyrrole units) with a central cobalt atom.

59. What are the coenzymes of vitamin B_{12}?

Vitamin B_{12} has two coenzymes — 5-deoxyadenosyl cobalamin and methylcobalamin.

60. What are the deficiency manifestations of vitamin B_{12}?

Causes pernicious anemia. It is characterized by decreased number of RBC, low hemoglobin levels and neurological manifestations.

61. Name the important vitamin like compounds.

Choline, inositol, lipoic acid, p-aminobenzoic acid.

62. What are the functions of choline?

Choline is a component of phospholipid, lecithin which is involved in membrane structure and function. Choline acts as a lipotropic factor (prevents the accumulation of fat in liver), and participates in one carbon metabolism.

63. What do you know about inositol?

It is a component of phosphatidyl- inositol (a phospholipid), acts as a lipotropic factor, participates as a second messenger in hormonal action.

64. What are the biochemical functions of lipoic acid?

Involved in the decarboxylation reactions. The conversion of pyruvate to acetyl CoA by pyruvate dehydrogenase and α-ketoglutarate to succinyl CoA by α-ketoglutarate dehydrogenase, require lipoic acid.

65. What all you know about para aminobenzoic acid?

Structural constituent of folic acid. Hence, PABA may be regarded as a vitamin in another vitamin (folic acid).

PHYSIOLOGICAL BIOCHEMISTRY II

8. Digestion and Absorption

MCQs

1. **The following foods need to be digested before absorption by the intestine except**
   (A) Polysaccharides and proteins
   (B) Amino acids and monosaccharides
   (C) Triacylglycerols and phospholipids
   (D) Nucleic acids and cholesteryl esters.
2. **The major function of liver with regard to digestion and absorption of foodstuffs**
   (A) Elaboration of proteolytic enzymes
   (B) Secretion of lipolytic enzymes
   (C) Synthesis of bile acids
   (D) Production of HCl.
3. **Salivary/pancreatic amylase specifically acts on**
   (A) α-1,4-glycosidic bonds
   (b) α-1,6-glycosidic bonds
   (C) β-1,4-glycosidic bonds
   (D) None of the above bonds.
4. **One of the following disaccharidase enzyme activity is low in a large number of adults**
   (A) Sucrase (B) Lactase
   (C) Maltase (D) Trehalase.
5. **The monosaccharide that is most rapidly absorbed in the small intestine**
   (A) Glucose (B) Fructose
   (C) Galactose (D) Arabinose.
6. **Glucose shares the same carrier mediated transport system as that of**
   (A) Na^+ (b) K^+ (C) Mg^{2+} (D) Ca^{2+}.
7. **The following statement(s) is (are) true about the intestinal absorption of glucose**
   (A) Absorption occurs against a concentration gradient
   (B) Energy is not directly utilized
   (C) Na^+ shares the same transport system
   (D) All of the above.
8. **One of the following polysaccharides is not digested by humans**
   (A) Glycogen (B) Starch
   (C) Cellulose (D) Dextrin.
9. **The problem of flatulence is primarily due to the non-digestible**
   (A) Oligosaccharides (B) Oligopeptides
   (C) Fats (D) RNAs.
10. **One of the following enzymes is absent in adults**
    (A) Renin (B) Rennin
    (C) Pepsin (D) Chymotrypsin.
11. **Trypsin cleaves the peptide bonds, the carbonyl (—CO—) group of which is contributed by**
    (A) Arginine and lysine
    (B) Arginine and histidine
    (C) Tyrosine and phenylalanine
    (D) Arginine and proline.
12. **Serine proteases (pepsin, trypsin) are inhibited by**
    (A) Allopurinol
    (B) Isonicotinic acid hydrazide
    (C) Methotrexate
    (D) Diisopropyl fluorophosphate.
13. **γ-Glutamyl cycle or Meister cycle involves a peptide namely**
    (A) Bradykinin (B) Glutathione
    (C) Enkephalin (D) Secretin.
14. **In Hartnup's disease, absorption of the following amino acid is severely affected**
    (A) Tyrosine (B) Phenylalanine
    (C) Tryptophan (D) Leucine.
15. **Hartnup's disease is characterized by the inability of the body to produce the following coenzyme(s)**
    (A) FMN and FAD (B) NAD^+ and $NADP^+$
    (C) Pyridoxal phosphate
    (D) Coenzyme A.
16. **About 90% of the lipid consumed by humans is in the form of**
    (A) Triacylglycerols (B) Free fatty acids
    (C) Cholesterol (D) Phospholipids.
17. **Emulsification of lipids is important for their digestion which is mainly brought out by**
    (A) Proteins (B) Phospholipids
    (C) Bile salts (D) Polysaccharides.
18. **The products of pancreatic lipase action are**
    (A) 1-monoacylglycerol + 2 free fatty acids
    (B) 2-monoacylglycerol + 2 free fatty acids
    (C) 3-monoacylglycerol + 2 free fatty acids
    (D) Glycerol + 3 free fatty acids.

19. The intestinal absorption of lipid components occurs by
(A) Diffusion (B) Facilitated diffusion
(C) Active transport (D) None of the above.

20. The major vehicles for intestinal absorption of lipids
(A) Chylomicrons (B) Liposomes
(C) Emulsions (D) Micelles.

21. The following type of lipoproteins are synthesized in the intestinal mucosal cells which enter the lymphatic system
(A) Chylomicrons
(B) Low density lipoproteins
(C) High density lipoproteins
(D) Very low density lipoproteins.

22. The dietary source of one of the following compounds does not significantly contribute to the body metabolism
(A) Monosaccharides and disaccharides
(B) Amino acids and peptides
(C) Cholesteryl esters
(D) Purines and pyrimidines.

Answers

1. **(B)** 2. **(C)** 3. **(A)** 4. **(B)** 5. **(C)** 6.**(A)** 7. **(D)** 8. **(C)** 9. **(A)** 10. **(B)** 11. **(A)** 12. **(D)** 13. **(B)** 14 **(C)**. 15. **(B)** 16. **(A)** 17. **(C)** 18. **(B)** 19. **(A)** 20. **(D)** 21. **(A)** 22. **(D)**.

Clinical Case Study 8.1
(Source : Biochemistry, Page 171)

A boy, aged 12 years complained abdominal discomfort and diarrhea with a feeling of being bloated, after consumption of milk. He was taken to a physician who advised him to stop his intake of milk. He felt better in 3 days.

1. **What was the disorder, the boy was suffering?**
2. **What is the explanation for diarrhea and flatulence?**

Ans. :

1. **Acquired lactose intolerance**.
2. The brush border of the intestine houses the enzyme lactase which is lost due to diarrhea, and hence typical symptoms of flatulence.

Clinical Case Study 8.2
(Source : Biochemistry, Page 173)

A 55-year old man had weight loss, diarrhea, abdominal pain, gas constipation. On consumption of wheat based diet to reduce his weight. The physician advised him to completely stop consumption of diet containing wheat or wheat products and switch over to the rice-based diet. The man had a remarkable improvement within a week with the new dietary regimen.

1. **What is the probable diagnosis?**
2. **Name nutrient linked with this disorder.**
3. **What happens to the absorption of other nutrients?**
4. **What is the most susceptible age for this?**

Ans. :

1. This is a case of **coeliac disease**.
2. Wheat proteins gluten/gliadin cause inflammation of intestinal wall.
3. The absorption of almost all the nutrients is impaired resulting in diarrhea, weight loss etc.
4. The incidence of coeliac disease is more common in children under 2 years age.

For related clinical case study, refer 6.3

VIVA/Short Questions

1. What do you understand by digestion?

The process involving the hydrolysis of large and complex organic molecules of foodstuffs into smaller and preferably water soluble molecules which can be easily absorbed by gastrointestinal tract.

2. How does cooking influence digestion?

Cooking significantly improves the digestibility of foodstuffs by the enzymes.

3. How do carbohydrates get digested in the small intestine?

The pancreatic α-amylase acts on starch to produce disaccharides (maltose, isomaltose) and oligosaccharides. The final digestion of di- and oligosaccharides is carried out by oligosaccharidases and disaccharidases (sucrase, lactase).

4. How are monosaccharides absorbed?

Among the monosaccharides, galactose is most efficiently absorbed followed by glucose and fructose.

5. What is the mechanism of absorption of glucose?

By a carrier mediated and energy requiring process. Glucose and Na^+ share the same carrier for transport. The enzyme Na^+-K^+ ATPase is involved in this transport of Na^+ in exchange for K^+.

6. What do you know about lactose intolerance?

Some individuals cannot tolerate the consumption of lactose due to the deficiency of the intestinal enzyme, lactase (β-galactosidase). The treatment involves elimination/restriction of lactose in the diet (milk and milk products).

7. What is flatulence?

Increased intestinal motility, cramps and irritation. This occurs due to the indigestion of certain carbohydrates (mostly non-digestible oligosaccharides such as raffinose of leguminous seeds).

8. Name the type of enzymes responsible for protein degradation?

Proteins are hydrolysed by peptidases.

9. What are the functions of gastric HCl?

Denaturation of proteins and killing of certain microorganisms. The denatured proteins are more susceptible for digestion.

10. How are proteins digested?

In the stomach by the action of pepsin, resulting in peptides and a few amino acids. Trypsin, chymotrypsin and elastase (of pancreatic juice) continue the protein digestion in the intestine to liberate oligopeptides and amino acids.

11. What do you know about the absorption of amino acids?

L-Amino acids are absorbed by an active process (against a concentration gradient). The mechanism of amino acid absorption is comparable to that of D-glucose. It is basically a Na^+-dependent active process and the energy is supplied by ATP.

12. What is Meister cycle?

Meister cycle or γ-glutamyl cycle is a type of amino acid transport system, involving a tripeptide namely glutathione.

13. What do you know about Hartnup's disease?

Hartnup is the name of the family in whom the disease was first discovered. It is characterized by the inability of intestinal and renal epithelial cells to absorb neutral amino acids. Tryptophan absorption is most severely affected with a result that typical symptoms of pellagra are observed in the patients of Hartnup's disease. This is related to the impairment in the conversion of tryptophan to NAD^+ and $NADP^+$, the coenzymes of niacin.

14. What is emulsification and how is it brought about?

It is a process of dispersion of lipids into smaller droplets due to reduction in the surface tension. Emulsification is brought about by 3 complementary mechanisms.

- Detergent action of bile salts (glycocholic acid and taurocholic acid).
- Surfactant action of degraded lipids.
- Mechanical mixing due to peristalsis.

15. How are fats digested?

Emulsification of fats is a prerequisite for the digestion. Pancreatic lipase is the major enzyme that digests fats (triacylglycerols).

16. What are the major abnormalities related to digestion and absorption?

Peptic ulcers, pancreatitis, lactose intolerance, deficiency of sucrase, Hartnup's disease and steatorrhea.

9. Plasma Proteins

MCQs

1. The packed cell volume (hematocrit) of human blood is around

(A) 35% (b) 45% (C) 55% (D) 65%.

2. The main difference between plasma and serum is the presence or absence of

(A) Fibrinogen (B) Albumin
(C) Globulins (D) Plasmin.

3. One electrophoresis, serum proteins can be separated into

(A) 3 bands (B) 5 bands
(C) 7 bands (D) 9 bands.

4. The normal albumin/globulin (A/G) ratio is 1.2 to 1.5 : 1. It may decrease in the following condition(s)

(A) Protein malnutrition (B) Renal damage
(C) Chronic infections (C) All of the above.

5. One of the following proteins is not synthesized in the liver

(A) Albumin (B) Transferrin
(C) Fibrinogen (D) Immunoglobulins.

6. The half-life of serum albumin is about

(A) 10 days (B) 20 days
(C) 30 days (D) 40 days.

7. The edema observed in kwashiorkor is attributed to a drastic reduction in the serum concentration of

(A) Albumin (B) Immunoglobulins
(C) Fibrinogen (D) Ceruloplasmin.

8. Plasma albumin performs the following function(s)

(A) Osmotic (B) Transport
(C) Nutritive (D) All of the above.

9. In heavy smokers, one of the amino acids of α_1-antitrypsin forms the following compound which in turn is implicated in emphysema
(A) Methionine oxide
(B) Methionine sulfoxide
(C) Cysteine sulfoxide
(D) Threonine sulfoxide.

10. Haptoglobin performs the following function
(A) Prevents the loss of hemoglobin through kidney
(B) Transports copper
(C) Participates in humoral immunity
(D) Promotes blood clotting.

11. In Wilson's disease, one of the following plasma protein concentration is lowered
(A) Haptoglobin (B) α_1-antitrypsin
(C) Ceruloplasmin (D) Transferrin.

12. Immunoglobins are synthesized by
(A) T-lymphocytes (B) Liver
(C) Kidney (D) B-lymphocytes.

13. On plasma (serum) electrophoresis the immunoglobins move along with the fraction
(A) α_1-globulins (B) α_2-globulins
(C) β-globulin (D) γ-globulins.

14. The structure of immunoglobulins basically consists of
(A) Two heavy chains and 2 light chains
(B) One heavy chain and 3 light chains
(C) Three heavy chains and 1 light chain
(D) Three heavy chains and 3 light chains.

15. The human immunoglobulins IgG, IgA, IgM, IgD and IgE respectively contain the following heavy chains
(A) α, γ, μ, δ and ε (B) γ, α, μ, δ and ε
(C) μ, δ, ε, α and γ (D) γ, μ, α, δ and ε.

16. The immunoglobulin that can cross the placenta and transfer the mother's immunity to the developing fetus
(A) IgG (B) IgA (C) IgM (D) IgD.

17. The following immunoglobulin is elevated in allergic reactions
(A) IgA (b) IgM (C) IgD (D) IgE.

18. Multiple myeloma is characterized by all of the following except
(A) Overproduction of mostly IgG
(B) Females are more susceptible than males
(C) It usually occurs in the age 45-60 years
(D) The body immunity remains normal.

19. Bence Jones proteins are the
(A) Heavy chains of immunoglobulins
(B) Light chains of immunoglobulins
(C) Both light and heavy chains
(D) Derivatives of albumin.

20. When acidified urine is heated to 40°-50°C, a precipitate is formed which redissolves on further heating. This indicates the presence of
(A) Albumin (B) Globulins
(C) Bence Jones proteins
(D) Immunoglobulin E.

21. The numbering (I, II, III......) of blood coagulation factors is based on the
(A) Order of their action
(B) Order of their discovery
(C) Increasing molecular weight
(D) Decreasing molecular weight.

22. One of the following has been deleted from the list of blood clotting factors
(A) IV (B) VI (C) IX (D) X.

23. The extrinsic and intrinsic pathways of blood clotting result in the activation of the factor
(A) IX (B) X (C) XI (D) XII.

24. Hemophilia is caused due to the deficiency of factor
(A) V (B) VI (C) VII (D) VIII.

25. Christmas disease is due to deficiency of factor
(A) VIII (B) IX (C) X (D) XI.

Answers

1. **(B)** 2. **(A)** 3. **(B)** 4. **(D)** 5. **(D)** 6. **(B)** 7. **(A)** 8. **(D)** 9. **(B)** 10. **(A)** 11. **(C)** 12. **(D)** 13. **(D)** 14. **(A)** 15. **(B)** 16. **(A)** 17. **(D)** 18. **(D)** 19. **(B)** 20. **(C)** 21. **(B)** 22. **(B)** 23. **(B)** 24. **(D)** 25. **(B)**.

Clinical Case Study 9.1
(Source : Biochemistry, Page 185)

A 30-year old man with a history of smoking for 15 years complained breathlessness and developed chronic bronchitis. His plasma concentration of α_1-antitrypsin was 20 mg/dl against the normal 200 mg/dl.

1. **What is the probable diagnosis?**
2. **How is a, α-antitrypsin deficiency linked to this disorder?**
3. **What is the biochemical basis of smoking involvement in this disorder?**

Ans. :

1. **Emphysema** (abnormal distension of lungs by air)
2. AT deficiency causes destruction of lung tissues due to lack of inhibition of the enzymes elastase and proteases in lungs.
3. Emphysema is more commonly associated with heavy smoking. This is because smoking causes oxidation of methionine to methionine sulfoxide, which cannot bind and inactivate proteases.

Clinical Case Study 9.2
(Source : Biochemistry, Page 192)

A 5-year old boy was hit by a motor vehicle in a road accident. It was observed that he had an unusual excessive bleeding. Laboratory investigations revealed that the boy had a prolonged clotting time and a very low concentration of clotting factor VIII.

1. **Describe the provisional diagnosis with justification.**
2. **What is the relation of this disorder with gender?**

Ans. :

1. The deficiency of clotting factor VIII (antihemophilic factor) and prolonged clotting time indicate that the boy was suffering from **classical hemophilia (hemophilia A)**.
2. This is a sex-linked disorder, transmitted by females affecting males.

For more clinical case studies, refer 20.6, 23.1, 23.2

VIVA/Short Questions

1. **What is the difference between plasma and serum?**

The main difference between plasma and serum is the presence or absence of fibrinogen. When blood containing anticoagulants (heparin, potassium oxalate) is centrifuged, the plasma separates out as a supernatant. Serum separates out when the blood clots.

2. **What are the important techniques to separate plasma proteins?**

By salting out technique (using ammonium sulfate or sodium sulfate), the plasma proteins can be separated into 3 groups—albumin, globulins and fibrinogen. Electrophoresis is the most commonly employed analytical technique for the separation of plasma (serum) proteins. Into 5 distinct bands namely albumin, α_1-, α_2-, β- and γ-globulins.

3. **What is the importance of serum electrophoresis pattern of proteins?**

Used for the diagnosis of certain diseases. e.g., in multiple myeloma, a sharp and distinct M-band appears in the γ-globulin fraction; in nephrotic syndrome, albumin is decreased with a prominent α_2-globulin; in primary immune deficiency, γ-globulin band is diminished.

4. **What is A/G ratio and what is its significance?**

Albumin/globulin (A/G) ratio expresses their relation in the serum concentration. The normal A/G ratio is 1.2 to 1.5 : 1. The A/G ratio is lowered either due to a decrease in albumin (liver diseases) or an increase in globulins (chronic infections).

5. **What do you know about plasma (serum) albumin?**

It is the major constituent of plasma proteins with a concentration 3.5-5.0 g/dl. It is exclusively synthesized by the liver. Plasma albumin performs osmotic, transport and nutritive function, besides the buffering action.

6. **What is the importance of α_1-antitrypsin?**

α_1-Antitrypsin (α_1-antiproteinase) is a component of γ-globulins. Its deficiency has been implicated in emphysema (abnormal distension of lungs).

7. **What all you know about haptoglobin?**

Haptoglobin (Hp) is a plasma glycoprotein. It binds with the free hemoglobin that spills into the plasma due to hemolysis. The haptoglobin-hemoglobin (Hp-Hb) complex cannot pass through glomeruli of kidney while free Hb can. Thus, Hp prevents the loss of Hb into the kidney. Hemolytic anemia is associated with decreased plasma concentration of haptoglobin.

8. **What is ceruloplasmin and what is its importance?**

Ceruloplasmin is a blue coloured, copper containing α_2-globulin. Its normal plasma concentration is around 30 mg/dl and it is decreased in Wilson's disease.

9. **What do you understand by immunity?**

The higher organisms including man have evolved a defense system collectively referred to as immunity.

(a) ***Cellular immunity*** mediated by T-lymphocytes (T-cells) that defends the body against the invasion of parasites, fungi, foreign tissues etc.

(b) ***Humoral immunity*** is mediated by immunoglobulins or antibodies. The B-lymphocytes or B-cells are responsible for the production of immunoglobulins.

10. What do you know about immunoglobulins?

Immunoglobulins (Ig) are a specialized group of glycoproteins synthesized by B-cells. The basic structure of an immunoglobulin consists of two identical heavy (H) chains and two identical light (L) chains held together by disulfide linkages.

11. How do you explain the production of unlimited number of antibodies?

The antibody diversity is achieved by two special processes namely combination of various structural genes and somatic mutations.

12. What is multiple myeloma?

A plasma cell cancer, constitutes about 1% of all cancers affecting population. This disease is characterized abnormal immunoglobulin production (mostly IgG).

13. What are Bence Jones proteins and how are they detected?

The light (κ and λ) chain immunoglobulins that are synthesized in excess. In about 20% of multiple myeloma patients, they are excreted in urine. Bence Jones proteins can be detected either by electrophoresis, heat test (i.e., at 40°-50°, Bence Jones protein precipitate and redissolve on further heating) and Bradshaw's test (addition of concentrated HCl to urine forms a white ring).

14. What do you know about the complement system?

It is composed of about 20 plasma proteins that complement the functions of antibodies in defending the body from invading antigens.

15. What do you understand by extrinsic and intrinsic pathways of blood clotting?

The extrinsic pathway refers to the process of clotting of factors that are not present in blood while intrinsic pathway involves the factors present in blood.

16. What do you know about anticoagulants?

Anticoagulants are the substances that inhibit blood clotting. These include oxalate, fluoride, EDTA and citrate, which can bind with Ca^{2+} and are very effective as anticoagulants. Heparin is a natural anticoagulant (secreted by mast cells) that maintains normal blood homeostasis.

17. What are the major abnormalities associated with blood clotting?

Hemophilia A (defect-antihemophilic factor i.e., VII), hemophilia B or Christmas disease (defect-Christmas factor, i.e., IX) are the major abnormalities known.

10. Hemoglobin and Porphyrins

MCQs

1. The following statements are true about hemoglobin except it

(A) Transports CO_2 and protons from tissues to lungs

(B) Is composed of two α and two β chains

(C) Is a tetrameric allosteric protein

(D) Acts as a reservoir for O_2 in skeletal muscle.

2. Iron of heme binds with the globin through the amino acid

(A) Histidine (B) Proline

(C) Tyrosine (D) Aspartate.

3. One of the following is important for the diagnosis and prognosis of diabetes mellitus

(A) HbA_1 (B) HbA_2 (C) HbA_{1c} (D) HbF.

4. One molecule of hemoglobin has

(A) Four hemes that can bind with 4 molecules of O_2

(B) Three hemes that can bind with 3 molecules of O_2

(C) Two hemes that can bind with 2 molecules of O_2

(D) Four hemes that can bind with 2 molecules of O_2.

5. Myoglobin has the following character(s).

(A) Hemoprotein found in heart and skeletal muscle

(B) Higher affinity for O_2 than hemoglobin

(C) Functions as a reservoir of oxygen

(D) All of the above.

6. One of the following enzymes of erythrocytes is closely associated with the transport of CO_2

(A) Pyruvate carboxylase

(B) Carbonic anhydrase

(C) Acetyl CoA carboxylase

(D) None of the above.

7. Bohr effect causes

(A) A shift in the oxygen dissociation curve to the right

(B) A shift in the oxygen dissociation curve to the left

(C) No change in the oxygen dissociation curve

(D) None of the above.

8. The following substance(s) interact(s) with oxyhemoglobin to release oxygen

(A) 2,3-bisphosphoglycerate
(B) CO_2 (C) H^+ and Cl^-
(D) All of these.

9. Methemoglobin is characterized by

(A) Methylation of heme
(B) Methylation of α-chain of globin
(C) Oxidation of Fe^{2+} to Fe^{3+} in hemoglobin
(D) Reduction of Fe^{3+} to Fe^{2+} in hemoglobin.

10. Sickle-cell anemia is characterized by replacement of

(A) Glutamate at 6th position of β-chain by valine
(B) Glutamate at 6th position of α-chain by valine
(C) Valine at 6th position of β-chain by glutamate
(D) Valine at 6th position of α-chain by glutamate.

11. Sickle-cell anemia patients exhibit the following abnormalities except

(A) Erythrocytes adopting a crescent like shape
(B) Resistance to general infections
(C) Life-long hemolytic anemia
(D) Premature death.

12. Sickle-cell trait provides resistance to malaria. This is explained by

(A) Shorter life span of erythrocytes
(B) The low pH of sickled cells
(C) The low concentration of K^+ in erythrocytes
(D) All of the above.

13. Thalassemias are characterized by a defect in the production of

(A) α-globin chain (B) β-globin chain
(C) α or β-globin chain (D) Heme.

14. The following proteins contain porphyrins except

(A) Myoglobin (B) Catalase
(C) Tryptophan pyrrolase (D) Fumarase.

15. One of the following intermediates of citric acid cycle is the immediate precursor for heme synthesis

(A) α-ketoglutarate (B) Succinyl CoA
(C) Oxaloacetate (D) Citrate.

16. The regulatory enzyme for heme biosynthesis in liver

(A) δ-aminolevulinate synthase
(B) δ-aminolevulinate dehydratase
(C) Heme synthase (D) Ferrochelatase.

17. The enzyme defect in acute intermittent porphyria

(A) Uroporphyrinogen I synthase
(B) Uroporphyrinogen III cosynthase
(C) Uroporphyrinogen decarboxylase
(D) Coproporphyrinogen oxidase.

18. The enzyme ferrochelatase is deficient in

(A) Hereditary coproporphyria
(B) Variegate porphyria (C) Protoporphyria
(D) Porphyria cutanea tarda.

19. The product of heme oxygenase reaction

(A) Bilirubin (B) Biliverdin
(C) Urobilin (D) Stercobilin.

20. Jaundice is a clinical condition characterized by yellow colour of the eyes (sclera) and skin. The term jaundice is derived from jaune which originated from

(A) Greek (B) Latin (C) Italian (D) French.

21. Criggler-Najjar syndrome is due to a defect in the enzyme

(A) UDP-glucuronyl transferase
(B) Heme oxygenase
(C) Biliverdin oxygenase
(D) None of the above.

22. The decision of insane King George (III) who ruled England) led to a war followed by American independance. It is widely believed that the mentally disturbed King was a victim of the following metabolic disorder.

(A) Alzheimer's disease
(B) Parkinson's disease
(C) Acute intermittent porphyria
(D) Alkaptonuria

23. Phototherapy is used for the newborns to treat neonatal physiologic jaundice. The following enzyme activity is low in the newborns.

(A) UDP-glucuronyltransferase
(B) Heme oxygenase (C) Biliverdin reductase
(D) Ferrochelatase.

Answers

1. **(D)** 2. **(A)** 3. **(C)** 4. **(A)** 5. **(D)** 6. **(B)** 7. **(A)** 8. **(D)** **9. (C) 10. (A) 11. (B) 12. (D) 13. (C) 14. (D) 15. (B)** 16. **(A)** 17. **(A)** 18. **(C)** 19. **(B)** 20. **(D)** 21. **(A)** 22. **(C)** 23. **(A)**.

Clinical Case Study 10.1
(Source : Biochemistry, Pages 210-211)

A 20 year old man had a history of cramping abdominal pain, diarrhaea, vomiting, tachycardia, insomnia, anxiety and muscle weakness, following intake of low-calorie diet for weight reduction. His laboratory investigative data are given hereunder.

Parameter	*Subject*	*Reference range*
Urinary porphobilinogen (PBG)	100 mg/24 hr	0-4 mg/24 hr
Urinary δ-aminolevulinate (ALA)	124 mg/24 hr	1-7 mg/24 hr

1. **What is the probable diagnosis?**
2. **Name the enzyme defect.**
3. **What is the influence of steroids and drugs?**
4. **What is the mode of treatment?**

Ans. :

1. Acute intermittent porphyria (AIP).
2. Uroporphyrinogen I synthase, and increased urinary excretion of porphobilinogen and δ-aminolevulinate.
3. AIP is precipitated by exogenous and endogenous factors such as steroids, porphyrinogenic drugs (e.g. barbiturates), alcohol, low calorie diets etc.
4. AIP can be treated by administration of hematin which inhibits the first enzyme of heme synthesis (δ-aminolevulinate synthase), and the accumulation of ALA and porphobilinogen.

Clinical Case Study 10.2
(Source : Biochemistry, Pages 201-203)

An African negro boy (aged 12 years), studying in an Indian High School, was admitted in a hospital with complaints of fever and severe pain in the arms and legs. His laboratory investigative data are given

Parameter	*Subject*	*Reference range*
Hemoglobin	7 g/dl	14-16 g/dl
Hematocrit	20 %	40-50 %
Hemoglobin electrophoresis	A distinct HbS band with slower movement than adult hemoglobin (HbA) detected	—
Microscopic examination of blood smear	Crescent (sickle) shaped erythrocytes	—

1. **What is the probable diagnosis?**
2. **What is the molecular basis of this disease?**
3. **The occurrence of this disorder provides resistant to malaria – explain.**

Ans. :

1. Sickle cell anemia
2. An abnormality in hemoglobin. It occurs due to the replacement of glutamate by valine at the 6th position of β-chain of hemoglobin.
3. Refer Chapter 10, page 204.

Clinical Case Study 10.3
(Source : Biochemistry, Page 216)

A 16-year old boy had viral infection. He had mild symptoms of jaundice — abdominal pain, anorexia and malaise. His laboratory findings

Parameter (serum)	*Subject*	*Reference range*
Bilirubin (total)	4.5 mg/dl	0.2-1 mg/dl
(van den Bergh reaction	Indirect positive	Negative)
Conjugated bilirubin	0.4 mg/dl	0.4 mg/dl
Unconjugated bilirubin	4.1 mg/dl	0.6 mg/dl

The above investigations were repeated six months later and the data obtained were almost the same as above.

1. **What is the probable diagnosis?**
2. **What is the enzyme defect?**
3. **How common is this disorder?**

Ans. :

1. Gilbert's syndrome
2. Hepatic UDP-glucuronyl transferase.
3. Gilbert's syndrome is more common in males and occurs in about 5% of the population.

Clinical Case Study 10.4
(Source : Biochemistry, Page 216)

A 3-day old male baby had yellow colouration of the sclerae of the eyes, and passed light yellow coloured urine. The serum total bilirubin concentration was 18 mg/dl (normal <1 mg/dl), most of it being unconjugated (van den Bergh indirect positive). The paediatrician advised phototherapy.

1. **What is the probable diagnosis?**
2. **What is the biochemical explanation for this disorder?**
3. **Which enzyme related to this disease is low in newborns?**
4. **What is the basis of phototherapy?**

Ans. :

1. **Physiological neonatal jaundice.**
2. Increased hemolysis coupled with immature hepatic system for the uptake, conjugation and secretion of bilirubin.
3. UDP-glucuronyl transferase.
4. The ultraviolet rays of the light isomerize bilirubin into a non-toxic form.

Clinical Case Study 10.5 (Source : Biochemistry, Page 201)

A child moved freely in a coal factory factory and inhaled smoke fumes. He developed lethargy and cyanosis. The spectrophotometric observations, of the blood showed absorption maxima at 500 nm and 631 nm.

1. **What is the probable diagnosis with justification?**
2. **What is the biochemical basis of this disorder?**

Ans. :

1. The child was exposed to industrial chemicals and the spectrophotometric analysis revealed the presence of methemoglobin (metHb). Here, the child suffered from toxic (acquired) **methemoglobinemia**.
2. Hemoglobin (Fe^{2+}) can be oxidized to methemoglobin (Fe^{3+}) by H_2O_2, by free radicals, drugs, industrial chemicals and smoke (aniline, nitro-benzene etc.). The methemoglobin cannot bind to O_2 and therefore O_2 supply to the tissues is impaired, hence the symptoms of lethargy and cyanosis.

Clinical Case Study 10.6 (Source : Biochemistry, Pages 205-206)

A one year old severely anemic male child was found to be totally lacking the machinery required for the synthesis of β-globin of hemoglobin. He was given frequent blood transfusions. However, he died at the age of 17 years.

1. **What is the provisional diagnosis?**
2. **What is the reason for this patient to survive upto 17 years against the usual death of patients around 2-3 years?**

Ans. :

1. **β-thalassemia major** due to a genetic defect in the synthesis of β-globin chain of hemoglobin.
2. The boy lived till the age of 17 years, thanks to repeated blood transfusions. Death might have occurred due to iron overload.

For more clinical case studies, refer 20.1, 20.4

VIVA/Short Questions

1. **What do you know about the structue of hemoglobin?**

It is a conjugated protein containing globin-the protein part, and the heme-the non-protein part (prosthetic group). The globin is composed of 2α and 2β chains. Heme contains protoporphyrin IX (with 4 pyrrole rings) with iron at the centre.

2. **What are the different types of hemoglobins?**

In adults, about 95% is hemoglobin A_1 (with 2α and 2β chains) and 5% hemoglobin A_2 (with 2α and 2δ chains).

3. **What do you understand by oxygen dissociation curve?**

The binding ability of hemoglobin with O_2 at different partial pressures of oxygen (pO_2) can be measured by a graphic representation known as oxygen dissociation curve.

4. **How does 2,3-bisphosphoglycerate (2-3, BPG) influence hemoglobin functions?**

It binds to deoxyhemoglobin and decreases the O_2 affinity to Hb. Further, the net effect of 2, 3-BPG is to facilitate the release of O_2 to the tissues.

5. **What is methemoglobin?**

Hemoglobin (Fe^{2+}) can be oxidized to methemoglobin (Fe^{3+}) by certain agent like H_2O_2, free radicals and drugs. The methemoglobin is unable to bind to O_2.

6. **What do you understand by hemoglobinopathies?**

The disorders caused by the synthesis of abnormal hemoglobin molecules or the production of insufficient quantities of normal hemoglobin or rarely both e.g., sickle-cell anemia, hemoglobin C disease.

7. **What is sickle-cell anemia (HbS)?**

The most common form of abnormal hemoglobins. The RBC of these patients adopt a sickle shape. In HbS, glutamate at sixth position of β-chain is replaced by valine. (Glu $\beta_6 \rightarrow$ Val). HbS is associated with hemolytic anemia, tissue damage, increased susceptibility to infections and premature death.

8. **How do you explain the resistance to malaria in sickle-cell anemia patients?**

The malarial parasite spends a part of its life cycle in erythrocytes. Increased lysis of sickle cells interrupts the parasite cycle. In addition, the concentration of K^+ is low in sickled cells which is unfavourable for the parasite to survive.

9. What do you understand by thalassemias?

A group of disorders characterized by impairment/imbalance in the synthesis of globin (α or β) chains of Hb e.g., α-thalassemia, β-thalassemia.

10. What are porphyrins?

Porphyrins are cyclic compounds composed of 4 pyrrole rings held together.

11. How is heme synthesis regulated?

The first committed step of heme biosynthesis catalysed by δ-aminolevulinate synthase is regulatory. It is inhibited through a feedback mechanism by heme.

12. What are porphyrias?

Porphyrias are the metabolic disorders of heme synthesis, characterized by the increased excretion of porphyrins or porphyrin precursors e.g., acute intermitent porphyria (ezyme defect-uroporphyrinogen I synthase); congenital erythropoietic porphyria (enzyme defect – uroporphyrinogen II cosynthase); variegate porphyria (enzyme defect-protoporphyrinogen oxidase).

13. How is heme degraded?

Heme is degraded to bile pigments, predominantly bilirubin (yellow coloured pigment).

14. How is bilirubin conjugated, and excreted into bile?

Bilirubin is conjugated with 2 molecules of glucuronate supplied by UDP-glucuronate. Bilirubin diglucuronide is excreted into the bile canaliculi against a concentration gradient by an active energy dependent process.

15. What is jaundice and what are different types?

Jaundice is a clinical condition characterized by yellow colour of the white of the eyes (sclerae) and skin. This occurs due to increased serum bilirubin concentration (normal <1 mg/dl).

There are mainly 3 types of jaundice.

(a) ***Hemolytic jaundice*** associated with increased hemolysis of erythrocytes.

(b) ***Hepatic jaundice*** due to impairment in liver function.

(c) ***Obstructive jaundice*** caused by an obstruction in the bile duct that prevents the passage of bile into the intestine.

16. What do you know about the jaundice due to genetic defects?

There are certain hereditary abnormalities that cause jaundice e.g., Crigler-Najjar syndrome type I and type II (enzyme defect-UDP-glucuronyltransferase).

11. Biological Oxidation

MCQs

1. The change in the heat content of a reaction is represented by

(A) Free energy (B) Enthalpy
(C) Entropy (D) Exergonic reaction.

2. All of the following are low energy compounds except

(A) Glucose 6-phosphate
(B) Carbamoyl phosphate
(C) ADP (D) Glycerol 3-phosphate.

3. All of the following are high energy compounds except

(A) Phosphoenolpyruvate
(B) Carbamoyl phosphate
(C) Creatine phosphate
(D) Fructose 6-phosphate.

4. One of the following is a high energy compound although it does not possess a phosphate group

(A) Acetyl CoA (B) AMP
(C) Glycerol (D) Uronic acid.

5. The storage form of high energy compound in invertebrates

(A) Creatine phosphate
(B) Arginine phosphate
(C) Carbamoyl phosphate
(D) Acetyl CoA.

6. The reducing equivalents of various metabolic intermediates are finally transferred to the following coenzymes for their entry into the electron transport chain

(A) FAD and FMN (B) NAD^+ and $NADP^+$
(C) FAD and NAD^+ (D) NAD^+ and FMN.

7. Mitochondrial matrix contains the enzymes responsible for

(A) Citric acid cycle (B) β-oxidation
(C) Oxidation of amino acids
(D) All of the above.

8. The mitochondrial matrix can be disrupted into five distinct complexes (I to V). One of them being an electron carrier, namely

(A) Complex I (B) Complex III
(C) Complex IV (D) Complex V.

9. The largest proportion of oxygen supplied to the body is utilized by the mitochondria for the operation of

(A) Citric acid cycle (B) Glycolysis
(C) β-oxidation
(D) Electron transport chain.

10. One of the following is not usually a substrate for electron transport chain

(A) $FADH_2$ (B) NADH
(C) NADPH (D) None of the above.

11. NAD^+ is reduced to NADH + H^+ by dehydrogenase from the following substrate(s)

(A) Glyceraldehyde 3-phosphate
(B) Pyruvate (C) Malate
(D) All of the above.

12. Isoprenoid side chain is present in one of the following components of electron transport chain

(A) Coenzyme Q (B) Cytochrome b
(C) Cytochrome a (D) Cytochrome oxidase.

13. The heme group of cytochromes differs from that found in the structure of hemoglobin in the character that the iron

(A) Is in ferrous (Fe^{2+}) state
(B) Is in the ferric (Fe^{3+}) state
(C) Gets alternately oxidized (Fe^{2+}) and reduced (Fe^{3+})
(D) Gets coordinated.

14. The P : O ratio for NADH

(A) 1.5 (B) 2.5 (C) 3 (D) 4.

15. Chemiosmotic hypothesis of oxidative phosphorylation explains ATP synthesis involving the translocation of

(A) Electrons (B) Protons
(C) Hydroxyl ions (D) Sodium ions.

16. The chemiosmotic hypothesis derives its strength from the following experimental observation(s)

(A) Inner mitochondrial membrane in impermeable to proton and other ions
(B) ATP can be generated by increasing the H^+ ion concentration on the outside of mitochondria
(C) ATP synthesis can be inhibited by compounds which increase the inner mitochondrial membrane permeability of protons
(D) All of the above.

17. Cytochrome oxidase of electron transport chain is inhibited by

(A) Carbon monoxide (B) Rotenone
(C) British antilewisite (D) Amytal.

18. The uncouplers do not allow one of the following to occur in the electron transport chain

(A) Oxidation of NADH and $FADH_2$
(B) Consumption of oxygen
(C) Formation of ATP
(D) Oxidation of only $FADH_2$.

19. One of the following hormones acts as an uncoupler

(A) Epinephrine (B) Thyroxine
(C) Insulin (D) Oxytocin.

20. The enzymes involved in biological oxidation come under the class of

(A) Oxidoreductases (B) Transferases
(C) Hydrolases (D) Ligases.

21. The following radical(s) obtained from oxygen is (are) toxic to the cell

(A) Superoxide (O_2^-)
(B) Hydrogen peroxide (H_2O_2)
(C) Hydroxyl (OH^-) (D) All of the above.

22. The following enzyme(s) is (are) involved in minimizing the toxic effects of free radicals

(A) Super oxide dismutase (B) Catalase
(C) Glutathione peroxidase
(D) All of the above.

23. The following protein, mostly found in the brown adipose tissue, acts as a physiological uncoupler

(A) NADH dehydrogenase
(B) Cytochrome oxidase
(C) Thermogenin (D) Thioredoxin

Answers

1. **(B)** 2. **(B)** 3. **(D)** 4. **(A)** 5. **(B)** 6. **(C)** 7. **(D)** 8. **(D)** 9. **(D)** 10. **(D)** 11. **(D)** 12. **(A)** 13. **(C)** 14. **(B)** 15. **(B)** 16. **(D)**. 17. **(A)** 18. **(C)** 19. **(B)** 20. **(A)** 21. **(D)** 22. **(D)** 23. **(C)**.

VIVA/Short Questions

1. What are high-energy compounds?

Substances which possess free energy to liberate at least 7 Cal/mol at pH 7.0. e.g., ATP, phosphoenol pyruvate, 1,3-bisphosphoglycerate, phosphocreatine.

2. 'ATP is the energy currency of the cell'—justify.

ATP can serve as an immediately available energy currency of the cell which is constantly being utilized and regenerated. ATP acts as an energy link between the catabolism and anabolism in biological process.

3. How is ATP synthesized?

ATP can be synthesized in two ways.

(a) ***Oxidative phosphorylation*** linked with electron transport chain is the major source of ATP synthesis.

(b) ***Substrate level phosphorylation*** involves the direct formation of ATP during substrate oxidation in the metabolism.

4. How do you define oxidation and reduction?

Oxidation is the loss of electrons, while reduction is the gain of electrons.

5. What is redox potential?

The oxidation-reduction (or redox) potential is a quantitative measure of the tendency of a redox pair to lose or gain electrons.

6. What is electron transport chain?

The reducing equivalents from various metabolic intermediates are transferred to form NADH and $FADH_2$. The latter two coenzymes pass through respiratory chain or electron transport chain (ETC) to finally reduce oxygen to water. ETC is located in the mitochondria.

7. 'Mitochondria are the power houses of cell'—Justify.

The mitochondria are the centres for metabolic oxidative reactions to generate reduced coenzymes (NADH and $FADH_2$) which, in turn, are utilized in ETC to liberate energy in the form of ATP. For this reason, mitochondria are appropriately regarded as the power houses of the cell.

8. What do you know about cytochromes?

Cytochromes are conjugated proteins containing heme. The electrons (in ETC) are transported from coenzyme Q to cytochromes (in the order) b, c_1, c and a_3.

9. What is cytochrome oxidase?

The term cytochrome oxidase is frequently used to collectively represent cytochrome a and a_3 which is the terminal component of ETC. Cytochrome oxidase is the only electron carrier, the heme iron of which can directly react with molecular oxygen.

10. What is oxidative phosphorylation?

The process of synthesizing ATP from ADP and Pi coupled with electron transport chain.

11. What do you understand by P : O ratio?

The atoms of phosphate utilized to the atoms of oxygen consumed in oxidation. P : O ratio represents the number of molecules of ATP synthesized for each pair of electrons carried through ETC. P : O ratio for NADH is 3, while that for $FADH_2$ is 2.

12. What do you know about the chemiosmotic hypothesis?

The translocation of protons (H^+) across the inner mitochondrial membrane creates a proton gradient between the matrix and intermembrane space. This proton gradient developed due to the electron flow in ETC is sufficient to result in the synthesis of ATP from ADP and Pi.

13. What are uncouplers?

Certain compounds uncouple (or delink) the electron transport chain from oxidative phosphorylation. The uncouplers allow oxidation of substrates (via NADH and $FADH_2$) without ATP formation e.g., dinitrophenol (DNP), dinitrocresol. The hormone thyroxine is a physiological uncoupler.

14. What is the significance of brown adipose tissue?

The mitochondria of these tissues are rich in electron carriers and are specialized to carry out oxidation uncoupled from oxidative phosphorylation. This causes liberation of heat when fat is oxidized in the brown adipose tissue. Thus, the presence of active brown adipose tissue in certain animals is believed to protect them from becoming obese.

15. What are ionophores?

The lipophilic substances that promote the transport of ions across biological membranes e.g., uncouplers (dinitrophenol, dinitrocresol), certain antibiotics (valinomycin, nigercin).

METABOLISMS III

12. Introduction to Metabolism

MCQs

1. The chemical reactions occurring in the living system are collectively referred to as

(A) Metabolism (B) Anabolism

(C) Catabolism (D) Energy metabolism.

2. One of the following processes is usually dependent on the supply of energy

(A) Catabolism (B) Anabolism

(C) Metabolism (D) None of the above.

3. The energy liberated during the metabolic reactions is primarily

(A) ADP (B) CTP (C) ATP (d) GTP

4. **Intermediary metabolism does not involve one of the following metabolisms**
   (A) Carbohydrates (B) Amino acids
   (C) Lipids (D) Nucleic acids.
5. **The reducing equivalents for the biosynthetic reactions are usually supplied by**
   (A) NADH (B) NADPH
   (C) $FADH_2$ (D) $FMNH_2$.
6. **The method(s) employed for the study of metabolism**
   (A) Whole organisms (B) Metabolic probes
   (C) Isotopes (D) All of the above.

Answers

1. (**A**) 2. (**B**) 3. (**C**) 4. (**D**) 5. (**B**) 6. (**D**).

For clinical case studies, refer Metabolism Chapters 13, 14, 15, 17

VIVA/Short Questions

1. **What do you understand by metabolism?**

The entire spectrum of chemical reactions, occurring in the living system.

2. **What is a metabolic pathway?**

A series of enzymatic reactions to produce specific products.

3. **Distinguish between catabolism and anabolism.**

Catabolism : The degradative processes concerned with the breakdown of complex molecules to simpler ones.

Anabolism : The biosynthetic reactions involving the formation of complex molecules from simple precursors.

4. **What is amphibolism?**

Amphibolism is used for reactions which are both catabolic and anabolic in nature.

5. **What do you understand by energy metabolism?**

The metabolic pathways concerned with the storage and liberation of energy.

6. **Name the different techniques used to study metabolic pathways/biochemical reactions?**

Use of whole organisms or its components, utility of metabolic probes, application of isotopes.

13. Metabolism of Carbohydrates

MCQs

1. **The monosaccharide that is the central molecule of carbohydrate metabolism**
   (A) Glucose (B) Fructose
   (C) Galactose (D) Ribose.
2. **The fasting blood glucose level in normal individuals**
   (A) 20-60 mg/dl (B) 60-100 mg/dl
   (C) 100-140 mg/dl (D) 140-200 mg/dl.
3. **Insulin is required for glucose uptake in the following organ/tissue**
   (A) Brain (B) Liver
   (C) Muscle (D) Erythrocytes.
4. **The enzymes of glycolysis are located in**
   (A) Mitochondria (B) Cytosol
   (C) Nucleus (D) Golgi apparatus.
5. **The end products of anaerobic glycolysis**
   (A) Pyruvate (B) Acetyl CoA
   (C) Lactate (D) Lactose.
6. **Glycolysis is of great importance for ATP production in**
   (A) Erythrocytes (B) Testes
   (C) Medulla of kidney (D) All of the above.
7. **The enzymes hexokinase and glucokinase differ from each other in the following character(s)**
   (A) K_m for substrate
   (B) Distribution in tissues
   (C) Inhibition by glucose 6-phosphate
   (D) All of the above.
8. **One of the following enzymes causes substrate level phosphorylation to generate ATP**
   (A) Hexokinase (B) Phosphofructokinase
   (C) Pyruvate kinase (D) Glucokinase.
9. **One of the following kinase reactions is reversible**
   (A) Hexokinase
   (B) Phosphofructokinase
   (C) Phosphoglycerate kinase
   (D) None of the above.
10. **The enzyme glyceraldehyde 3-phosphate dehydrogenase is inhibited by**
    (A) Arsenite (B) Arsenate
    (C) Fluoride (D) Oxamate.

11. For glucose estimation in the laboratory, blood is collected in a fluoride bulb. Fluoride inhibits the following enzyme in glycolysis

(A) Hexokinase (B) Aldolase
(C) Enolase (D) Pyruvate kinase.

12. Under anaerobic conditions, the number of ATP synthesized in glycolysis

(A) 1 (B) 2 (C) 4 (d) 8.

13. The most important regulatory enzyme catalysing the committed step in glycolysis

(A) Hexokinase
(B) Phosphofructokinase
(C) Pyruvate kinase (D) Enolase.

14. One of the following enzymes in glycolysis is a bifunctional enzyme

(A) Pyruvate kinase (B) Enolase
(C) Phosphofructokinase (D) Hexokinase.

15. Rapaport-Leubering cycle, operative in RBC, is mainly concerned with the production of

(A) 1,3-bisphosphoglycerate
(B) 2,3-bisphosphoglycerate
(C) Pyruvate (D) Lactate.

16. Synthesis of 2, 3-bisphosphoglycerate in erythrocytes is linked with

(A) Transport of CO_2 to lungs
(B) Release of O_2 to the tissues
(C) Maintenance of pH
(D) Citric acid cycle.

17. Pyruvate dehydrogenase complex requires the following coenzymes

(A) TPP, lipoamide, FAD, coenzyme A, $NADP^+$
(B) TPP, lipoamide, FAD, biotin, NAD^+
(C) TPP, lipoamide, FMN, coenzyme A, NAD^+
(D) TPP, lipoamide, FAD, coenzyme A, NAD^+.

18. A comparable enzyme with pyruvate dehydrogenase in the citric acid cycle

(A) Isocitrate dehydrogenase
(B) α-ketoglutarate dehydrogenase
(C) Succinate dehydrogenase
(D) Malate dehydrogenase.

19. The final common oxidative pathway for the foodstuffs

(A) Glycolysis (B) HMP shunt
(C) Citric acid cycle (D) β-oxidation.

20. One of the following metabolites is considered to play a catalytic role in citric acid cycle

(A) Citrate (B) Acetyl CoA
(C) Oxaloacetate (D) Oxalosuccinate.

21. One of the following metabolites of citric acid cycle serves as a good source for cytosolic acetyl CoA

(A) Citrate (B) Isocitrate
(C) α-ketoglutarate (D) Oxaloacetate.

22. The number of $NADH+H^+$ produced in citric acid cycle

(A) 1 (B) 2 (C) 3 (d) 4.

23. Citric acid cycle operates only in the presence of oxygen because O_2

(A) Is consumed in cycle proper
(B) Is used in electron transport chain for the regeneration of NAD^+ and FAD
(C) Is essential for the formation of acetyl CoA
(D) None of the above.

24. The following reaction(s) is (are) considered anaplerotic in nature

(A) Conversion of pyruvate to oxaloacetate
(B) Synthesis of malate from pyruvate
(C) Formation of α-ketoglutarate from glutamate
(D) All of the above.

25. When a molecule of glucose is completely oxidized, the number of ATP generated

(A) 2 (B) 8 (C) 38 (D) 32.

26. The following are good substrates for gluconeogenesis

(A) Lactate, pyruvate (B) Alanine, tyrosine
(C) Propionate, glycerol
(D) All of the above.

27. More than 60% of the body glucose at rest is consumed by

(A) Skeletal muscle (B) Liver
(C) Brain (D) Adipose tissue.

28. Under anaerobic conditions, the energy source for skeletal muscle

(A) Glucose (B) Fatty acids
(C) Glycerol (D) Amino acids.

29. The following enzyme(s) is (are) involved in gluconeogenesis

(A) Pyruvate carboxylase
(B) Fructose 1,6-bisphosphatase
(C) Glucose 6-phosphatase
(D) All of the above.

30. All of the following amino acids are glucogenic except

(A) Alanine, serine
(B) Asparate, glutamate
(C) Histidine, methionine
(D) Leucine, lysine.

31. The enzyme glycerokinase is absent in
(A) Liver (B) Kidney
(C) Adipose tissue (D) None of the above.

32. The metabolite that is regarded as the dead end in glycolysis
(A) Pyruvate (B) Lactate
(C) 2,3-bisphosphoglycerate
(D) 3-phosphoglycerate.

33. The amino acid that is predominantly transported from skeletal muscle to liver for gluconeogenesis
(A) Glycine (B) Alanine
(C) Aspartate (D) Tryptophan.

34. One of the following derivatives of fat is a precursor for gluconeogenesis
(A) Fatty acids (B) Acetyl CoA
(C) Propionyl CoA
(D) None of the above.

35. The following organ(s) is (are) dependent on the continuous supply of glucose for energy
(A) Brain (B) Erythrocytes
(C) Testes (D) All of the above.

36. The largest quantity of glycogen is present in
(A) Liver (B) Muscle (C) Brain (D) Kidney.

37. Glycogenin is the primer required for glycogen synthesis in the absence of a glycogen primer. The site at which glucose binds to glycogenin is the hydroxyl group of
(A) Tyrosine (B) Serine
(C) Threonine (D) Hydroxyproline.

38. Glycogen synthase catalyses the formation of
(A) α 1,4-glycosidic bonds
(B) α 1,6-glycosidic bonds
(C) β 1,4-glycosidic bonds
(D) None of the above.

39. Glycogen phosphorylase acts on glycogen to produce
(A) Glucose 6-phosphate
(B) Glucose 1-phosphate
(C) Free glucose
(D) Fructose 1-phosphate.

40. One of the following enzymes of glycogen metabolism contains pyridoxal phosphate
(A) Glycogen synthase
(B) Glycogen phosphorylase
(C) Phosphoglucomutase
(D) Debranching enzyme.

41. By the combined action of glycogen phosphorylase and debranching enzyme on glycogen, the formation of glucose 1-phosphate and free glucose is in the ratio of
(A) 2 : 1 (B) 4 : 1 (C) 8 : 1 (d) 16 : 1.

42. The enzyme glucose 6-phosphatase is present in the following organs except
(A) Liver (B) Kidney
(C) Intestine (D) Muscle.

43. Not a single molecule of free glucose is liberated through glycogenolysis in the following organ
(A) Liver (B) Kidney
(C) Muscle (D) None of the above.

44. In a well-fed state, one of the following allosterically activates the enzyme glycogen synthase
(A) ATP (B) Glucose
(C) Glucose 6-phosphate (D) Glycogen.

45. The enzyme glycogen phosphorylase is allosterically inhibited by
(A) ATP (B) Glucose
(C) Glucose 6-phosphate
(D) All of the above.

46. Through the activities of cyclic AMP, the following hormone(s) inhibit(s) glycogenesis
(A) Glucagon (B) Epinephrine
(C) Norepinephrine (D) All of the above.

47. One of the following hormones increases glycogenesis
(A) Glucagon (B) Insulin
(C) Epinephrine (D) Norepinephrine.

48. The enzyme defect in von Gierke's disease (type I glycogen storage disease)
(A) Glucose 6-phosphatase
(B) α 1,4-glucosidase
(C) Debranching enzyme
(D) Phosphofructokinase.

49. Hexose monophosphate shunt is characterized by all of the following except
(A) Direct oxidation of glucose to CO_2
(B) Occurs in cytosomal fraction of the cell
(C) Produces NADPH and pentoses
(D) Generates 38 ATP.

50. Pentoses produced in HMP shunt are utilized for the synthesis of
(A) DNA and RNA (B) NAD^+ and $NADP^+$
(C) FAD and coenzyme A
(D) All of the above.

51. The generation of NADPH in hexose monophosphate shunt is associated with the activity of the enzyme
(A) Glucose 6-phosphate dehydrogenase
(B) Transketolase (C) Transaldolase
(D) None of the above.

52. Transketolase is dependent on the coenzyme
(A) Thiamine pyrophosphate
(B) Pyridoxal phosphate
(C) $NADP^+$ (D) Coenzyme A.

53. Glucose 6-phosphate dehydrogenase in RBC is associated with the maintenance of
(A) Reduced glutathione
(B) Integrity of membrane structure
(C) Ferrous (Fe^{2+}) form of iron in hemoglobin
(D) All of the above.

54. The victims of glucose 6-phosphate dehydrogenase deficiency are resistant to
(A) Malaria (B) Filaria
(C) Tuberculosis (D) Hepatitis.

55. Uronic acid pathway is concerned with the formation of
(A) Glucuronic acid (B) Pentoses
(C) Vitamin C (D) All of the above.

56. Man and some other animals cannot synthesize ascorbic acid due the deficiency of the enzyme
(A) Glucuronidase
(B) UDP-glucose dehydrogenase
(C) L-gulonolactone oxidase
(D) Xylitol dehydrogenase.

57. One of the following monosaccharides is an essential nutrient for humans
(A) Glucose (B) Fructose
(C) Galactose (D) None of the above.

58. The enzyme defect in galactosemia
(A) Galactose 1-phosphate uridyltransferase
(B) UDP-hexose epimerase
(C) Aldolase B
(D) L-gulonolactone oxidase.

59. The following metabolite formed from galactose has been implicated in the development of cataract
(A) Sorbitol (B) Galactitol
(C) UDP-galactose
(D) Galactose 1-phosphate

60. The pathological changes observed in diabetes mellitus are believed to be due to the accumulation of the following metabolite
(A) Galactitol (B) Sorbitol
(C) Fructose 1-phosphate (D) Glycerol.

61. Hereditary fructose intolerance is due to the deficiency of the enzyme
(A) Aldolase (B) Aldolase B
(C) Fructokinase (D) Triokinase.

62. The preferred carbohydrate source of energy by sperms
(A) Glucose (B) Galactose
(C) Ribose (D) Fructose.

63. Plants and microorganisms can convert fat to glucose via
(A) Gluconeogenesis (B) Glyoxylate cycle
(C) Krebs cycle (D) Glycolysis.

64. In glyoxylate cycle, the enzyme that cleaves isocitrate to glyoxylate and succinate
(A) Isocitrate lyase (B) Isocitrate synthase
(C) Citrate synthase (D) Aconitase.

65. A 52 year old male, with a history of type 2 diabetes mellitus, on treatment with drugs and Insulin went to a remote village and returned after 3 months. He consulted his physician for routine health check up. Which investigation is advised to know his glycemic status during the last 2-3 months?
(A) Fasting blood glucose
(B) Random blood glucose
(C) Glycosylated hemoglobin
(D) Fructosamine.

66. A 45 year old male, disturbed in personal life, was consuming alcohol for 2 days without taking proper food. Laboratory results showed he was hypoglycemic. Which carbohydrate metabolism is impaired?
(A) Glycogenesis (B) Gluconeogenesis
(C) Glycolysis (D) Glycogenolysis.

67. On routine examination, a 5 year old boy's urine was positive for Benedict's test. However, by glucose oxidase method, it was negative. Further investigations revealed that the boy was a victim of essential pentosuria. Name the defective enzyme in this disorder.
(A) Galactose 1-phosphate uridyltransferase
(B) L-Gulonolactone oxidase
(C) Glycogen synthase
(D) Xylitol dehydrogenase

Answers

1. **(A)** 2. **(B)** 3. **(C)** 4. **(B)** 5. **(C)** 6. **(D)**. 7. **(D)** 8. **(C)** 9. **(C)** 10. **(B)** 11. **(C)** 12. **(B)** 13. **(B)** 14. **(C)** 15. **(B)** 16. **(B)** 17. **(D)** 18. **(B)** 19. **(C)** 20. **(C)** 21. **(A)** 22. **(C)** 23. **(B)** 24. **(D)** 25. **(D)** 26. **(D)** 27. **(C)** 28. **(A)** 29. **(D)** 30. **(D)** 31. **(C)** 32. **(B)** 33. **(B)** 34. **(C)** 35. **(D)** 36. **(B)** 37. **(A)** 38. **(A)** 39. **(B)** 40. **(B)** 41. **(C)** 42. **(D)** 43. **(D)** 44. **(C)** 45. **(C)** 46. **(D)** 47. **(B)** 48. **(A)** 49. **(D)** 50. **(D)** 51. **(A)** 52. **(A)** 53. **(D)** 54. **(A)** 55. **(D)** 56. **(C)** 57. **(D)** 58. **(A)** 59. **(B)** 60. **(B)** 61. **(B)** 62. **(D)** 63. **(B)** 64. **(A)** 65. **(C)** 66. **(B)** 67. **(D)**.

Clinical Case Study 13.1
(Source : Biochemistry, Page 279)

A normal one month old baby had a history of vomiting and diarrhea that frequently occurred after breast feeding. The urine gave a positive test for reducing sugars Benedict's test while the (specific for glucose). The RBC were found to be totally deprived of activity of the enzyme galactose 1-phosphate uridyltransferase.

1. **What is the probable diagnosis? Name the enzyme defect.**
2. **Which compound is elevated and excreted into urine?**
3. **What is the nutritional advice to overcome this disorder?**

Ans. :

1. **Galactosemia**, a metabolic disorder, due to the deficiency of the enzyme galactose 1-phosphate uridyltransferase.
2. Increase in blood galactose level, and its excretion into urine.
3. For the babies suffering from galactosemia, milk has to be removed from the diet and replaced with infant formula containing sucrose.

Clinical Case Study 13.2
(Source : Biochemistry, Page 269)

An 8-year old girl with enlarged abdomen was irritable and lethargic. She often had the symptoms of weakness and sweating which disappeared on eating. The girl was found to have enlarged liver. The laboratory data of the subject

Parameter	*Subject*	*Reference range*
Blood glucose (fasting)	40 mg/dl	70-100 mg/dl
Blood pH	7.25	7.35-7.45
Blood ketone bodies (total)	5 mg/dl	2-3 mg/dl
Serum lactate	25 mg/dl	3-8 mg/dl
Serum triglycerides	350 mg/dl	75-150 mg/dl
Serum uric acid	10 mg/dl	4-6 mg/di

1. **What is the probable diagnosis?**
2. **Name the enzyme defect.**
3. **What is the reason for the frequent episodes of hypoglycemia?**
4. **What is the explanation for elevated TG, ketone bodies, uric acid etc?**

Ans. :

1. **von Gierke's disease (glycogen storage disease type I).**
2. Glucose 6-phosphatase in liver.
3. Enough free glucose is not liberated in the liver to maintain blood glucose level. Hence the girl had frequent episodes of hypoglycemia.
4. The substrate glucose 6-phosphate is diverted to other pathways (HMP shunt, lipogenesis etc.) ultimately resulting in elevated triglycerides, ketone bodies, uric acid etc.

Clinical Case Study 13.3
(Source : Biochemistry, Page 275)

A 25-year old man was on treatment with antimalarial drug primaquine. As the treatment was in progress, he developed complications. This subject's laboratory investigative data are given

Parameter	*Subject*	*Reference range*
Hematocrit	30 %	45-50 %
Hemoglobin		
before treatment	12 g/dl	12-14 g/dl
after treatment	8 g/dl	
Serum bilirubin	3.5 mg/dl	< 1.0 mg/dl

The laboratory tests also indicated increased fragility of erythrocytes.

1. **What is the probable diagnosis?**
2. **How does the drug primaquine affect RBC?**
3. **What is the explanation for reduced hemoglobin and increased bilirubin?**

Ans. :

1. **Deficiency of the enzyme glucose 6-phosphate dehydrogenase (G6PD).**
2. The drugs such as primaquine cause increased generation of free radicals which cannot be effectively inactivated due to G6PD deficiency.
3. Hemolysis of RBC and consequently jaundice, as is evident from the laboratory data-reduced hemoglobin and increased serum bilirubin.

Clinical Case Study 13.4
(Source : Biochemistry, Pages 262-263)

A 35-year old chronic alcoholic man, had palpitations, severe sweating, confusion and fainted. His wife, fearing that her husband suffered heart attack, immediately rushed him to the nearby cardiac centre. All the tests indicative of myocardial infarction were negative. On enquiry, his wife informed the physician that he had not taken breakfast. The following are the laboratory data

Parameter	*Subject*	*Reference range*
Blood glucose	45 mg/dl	70-100 mg/dl
Serum lactate	15 mg/dl	0.3-7 mg/dl

The man was given intravenous glucose and responded immediately, and gained consciousness.

1. **What is the probable diagnosis?**
2. **What is the biochemical explanation for this disorder?**

Ans. :

1. **Hypoglycemia due to alcohol consumption**.
2. Gluconeogenesis is reduced due to alcohol intake. This is so since more of NADH is produced in alcohol metabolism which converts pyruvate to lactate. This limits pyruvate conversion to glucose, a major pathway of gluconeogenesis.

For more clinical case studies, refer 7.4, 16.1, 19.1, 21.1, 21.2

VIVA/Short Questions

1. **What is the central molecule of carbohydrate metabolism?**

Glucose.

2. **What is the fasting blood glucose level in normal individuals?**

70-100 mg/dl (4.5-5.5 mmol/l)

3. **Name the cells/organs where insulin–dependent transport system of glucose occurs.**

Mescle, adipose tissue.

4. **Name the cells/organs where insulin–independent transport system of glucose occurs.**

Hepatocytes, erythrocytes and brain.

5. **What is glycolysis.**

The sequence of reactions converting glucose (or glycogen) to pyruvate or lactate, with the production of ATP.

6. **Name the cells/tissues lacking mitochondria where glycolysis is the major pathway for ATP synthesis.**

Erythrocytes, cornea, lens etc.

7. **In which cellular part the enzymes of glycolysis are located?**

Cytosomal fraction.

8. **What are the products formed in glycolysis under aerobic and anaerobic conditions.**

Pyruvate – aerobic; lactate – anaerobic

9. **What are the differences between hexokinase and glucokinase?**

Hexokinase is present in almost all the tissues. It catalyses the phosphorylation of various hexoses (fructose, mannose etc.), has low K_m for substrates (about 0.1 mM) and is inhibited by glucose 6-phosphate.

10. **Name the enzymes involved in substrate level phosphorylation in glycolysis.**

1,3-bisphosphoglycerate kinase, and pyruvate kinase.

11. **What is the significance of conversion of pyruvate to lactate?**

The formation of lactate allows the regeneration of NAD^+ which can be reused by glyceraldehyde 3-phosphate dehydrogenase so that glycolysis proceeds even in the absence of oxygen to supply ATP.

Under anaerobic conditions, 2 ATP are synthesized while, under aerobic conditions, 7 ATP.

12. **Name the enzyme inhibited by fluoride in glycolysis. What is its significance.**

Fluoride; useful for correct estimation of blood glucose in the laboratory.

13. **What is Rapoport-Leubering cycle?**

Supplementary pathway to glycolysis concerned with the synthesis of 2,3-bisphosphoglycerate (2,3-BPG) in the RBC.

14. **What is the importance of 2,3-BPG?**

In the presence of 2,3-BPG, oxyhemoglobin unloads more oxygen to the tissues.

15. **Name enzyme and coenzymes for the conversion of pyruvate to acetyl CoA.**

Pyruvate dehydrogenase complex (PDH), TPP, lipoamide, FAD, coenzyme A and NAD^+.

16. **What is TCA cycle?**

TCA cycle (Krebs cycle or Citric acid cycle) essentially involves the oxidation of acetyl CoA to CO_2 and H_2O. It is the final common oxidative pathway for carbohydrates, fats and amino acids.

17. **Where are the enzymes of Krebs cycle located?**

The enzymes of TCA cycle are located in mitochondrial matrix.

18. TCA cycle is not a closed cycle - Justify.

TCA cycle—an open cycle since many compounds enter the cycle and leave.

19. Name the enzyme and coenzymes for the conversion of α-ketoglutarate to succinyl CoA.

α-ketoglutarate dehydrogenase complex. TPP, lipoamide, NAD^+, FAD and CoA.

20. Where does substrate level phosphorylation occur in TCA cycle?

Conversion of succinyl CoA to succinate by succinate thiokinase (GTP synthesized).

21. How many ATP are produced when one acetyl CoA oxidized in TCA cycle.

Ten

22. How many ATP are generated when a molecule of glucose is metabolized?

Under aerobic condition - 32; Under anaerobic condition - 2.

23. Why Krebs cycle is regarded as amphibolic in nature?

Krebs cycle is both catabolic and anabolic in nature, hence regarded as amphibolic.

24. What do you understand by anaplerosis?

The reactions concerned to replenish or to fill up the intermediates of citric acid cycle are called anaplerotic reactions or anaplerosis.

25. What is gluconeogenesis, and what are the major substrates?

Synthesis of glucose from non-carbohydrate compounds, lactate, pyruvate, glucogenic amino acids, propionate and glycerol.

26. Name the three irreversible enzymes in glycolysis where alternate arrangements are made for gluconeogenesis?

Hexokinase (or glucokinase), phosphofructokinase, pyruvate kinase.

27. Which is the metabolite formed by the oxidation of odd chain fatty acids that can form glucose.

Propionyl CoA

28. What is Cori cycle.

Synthesis of glucose in liver from the skeletal muscle lactate and the reuse of glucose thus synthesized by the muscle for energy purpose.

29. How is gluconeogenesis influenced by glucagon and insulin?

Glucagon promotes while insulin inhibits gluconeogenesis.

30. What is the effect of alcohol consumption on gluconeogenesis?

Reduces gluconeogenesis thereby increasing the risk of hypoglycemia.

31. What are substrates for gluconeogenesis, derived from fat?

Glycerol and propionate.

32. What is glycogenesis?

Synthesis of glycogen from glucose is glycogenesis.

33. What is glycogenin?

A protein primer to initiate the synthesis of glycogen.

34. What is glycogenolysis?

Degradation of stored glycogen in liver and muscle constitutes glycogenolysis.

35. Name the tissues which cannot produce free glucose due to absent of the enzyme glucose 6-phosphatase.

Muscle and brain.

36. Name the regulatory enzymes of glycogenesis and glycogenolysis.

Glycogen synthase - glycogenesis

Glycogen phosphorylase - glycogenolysis

37. What is the overall effect of hormones on glycogen metabolism?

An elevated glucagon or epinephrine level increases glycogen degradation whereas an elevated insulin results in increased glycogen synthesis.

38. What are futile cycles?

The synthesis and degradative metabolic pathways of a particular substance (say gluconeogenesis and glycolysis related to glucose) operate to the same extent simultaneously, this would result in futile cycles.

39. Name in enzyme defect and characteristic features of von Gierke's disease.

Glucose 6-phosphatase; fasting hypoglycemia, lactic acidemia, hyperlipidemia, hyperuricemia, gouty arthritis.

40. What is HMP shunt?

An alternative pathway to glycolysis and TCA cycle for the oxidation of glucose.

41. What is the regulatory enzyme of HMP shunt?

Glucose 6-phosphate dehydrogenase (G6PD).

42. Name the coenzyme required by transketolase.

Thiamine pyrophosphate (TPP).

43. What are the two metabolically important products, produced in HMP shunt?

Pentoses and NADPH.

44. What is the importance of pentoses?

Synthesis of nucleic acids (RNA and DNA) and many nucleotides such as ATP, NAD^+, FAD and CoA.

45. What is the significance of NADPH?

Biosynthesis of fatty acids and steroids, synthesis of certain amino acids, antioxidant (free radical scavenging) reactions, detoxification of drugs, phagocytosis, integrity of RBC membrane.

46. What is explanation for G6PD deficient people's resistance to malaria?

Parasites that cause malaria are dependent on HMP shunt and reduced glutathione for their optimum growth in RBC. Therefore, G6PD deficiency—protects them from malaria.

47. What is the importance of uronic acid pathway?

Synthesis of glucuronic acid, pentoses, and vitamin – ascorbic acid (except in primates and guinea pigs).

48. Name the enzyme deficient in humans due to which ascorbic acid cannot be synthesized?

L-gulonolactone oxidase—which converts gulonate to ascorbic acid.

49. What is essential pentosuria?

Deficiency of xylitol dehydrogenase. The affected individuals excrete large amounts of L-xylulose in urine.

50. What is galactosemia, and what are its manifestations?

Due to the deficiency of the enzyme galactose 1-phosphate uridyltransferase. Increased galactose levels in circulation (galactosemia) and urine (galactosuria). Implicated in the development of cataract.

51. Name the compound produced in galactose metabolism implicated in the development of cataract.

Galactitol (dulcitol)

52. What is sorbitol/polyol pathway?

Involves the conversion of glucose to fructose via sorbitol.

53. What is the danger of sorbitol accumulation in diabetes mellitus.

Some of the pathological changes associated with diabetes (like cataract formation, peripheral neuropathy, nephropathy etc.) are believed to be due to the accumulation of sorbitol.

54. What do you know about essential fructosuria?

Deficiency of the enzyme hepatic fructokinase. With excretion of fructose in urine. Treatment involves the restriction of dietary fructose.

55. What are mucopolysaccharidoses?

Lysosomal storage diseases caused by enzyme defects in the degradation of glycosaminoglycans.

56. What is glyoxylate cycle?

Plants and many microorganisms convert fat into carbohydrates by glyoxylate cycle.

14. Metabolism of Lipids

MCQs

1. The most abundant lipids (comprising 80-90% of the total) of the body

(A) Phospholipids (B) Triacylglycerols
(C) Lipoproteins (D) Cholesterol.

2. The hormone sensitive triacylglycerol lipase is activated through cyclic AMP by the following hormones except

(A) Epinephrine (B) Glucagon
(C) Thyroxine (D) Insulin.

3. The free fatty acids released by lipolysis in the adipose tissue are transported in the plasma in a bound form to

(A) Low density lipoproteins
(B) Chylomicrons (C) Albumin
(D) High density lipoproteins.

4. The fatty acids cannot be oxidized by one of the following tissues/organs

(A) Liver (B) Adipose tissue
(C) Muscle (D) Brain.

5. The fatty acids in the body are predominantly oxidized by

(A) α-oxidation (B) β-oxidation
(C) γ-oxidation (D) ω-oxidation.

6. One of the following compounds is involved in the transport of activated fatty acids from the cytosol to mitochondria

(A) Creatine (B) Creatinine
(C) Carnitine (D) Lysine.

7. The number of high energy phosphates utilized for the activation of fatty acids by the enzyme thiokinase

(A) 1 (B) 2 (C) 3 (D) 4.

8. One of the following compounds strongly inhibits the carnitine transport system of long chain fatty acids

(A) Acetyl CoA (B) Malonyl CoA
(C) Acyl CoA (D) Coenzyme A.

9. β-Oxidation is associated with a sequential removal of

(A) Acetyl CoA (B) Malonyl CoA
(C) Propionyl CoA (D) Butyryl CoA.

10. The number of acetyl CoA molecules liberated in the oxidation of palmitic acid

(A) 6 (B) 7 (C) 8 (D) 9.

11. The total number of ATP synthesized in the complete oxidation of palmitic acid

(A) 38 (B) 36 (C) 5 (D) 106.

12. The final products obtained in the oxidation of odd chain carbon containing fatty acids

(A) 2 molecules of acetyl CoA
(B) 1 molecule of each of acetyl CoA and propionyl CoA
(C) 1 molecule each of acetyl CoA and malonyl CoA
(D) 1 molecule each of propionyl CoA and malonyl CoA.

13. The conversion of propionyl CoA to succinyl CoA requires the following two coenzymes/ vitamins

(A) Thiamine pyrophosphate and biotin
(B) FAD and NAD^+
(C) Biotin and vitamin B_{12}
(D) Pantothenic acid and lipoic acid

14. β-Oxidation of fatty acids in peroxisomes is characterized by the following except

(A) ATP is synthesized
(B) $FADH_2$ oxidized to FAD and H_2O_2
(C) Reducing equivalents do not pass through electron transport chain
(D) Hypolipidemic drug induced peroxisomal oxidation.

15. The disorder characterized by the absence of peroxisomes in the tissues

(A) Refsum's disease (B) Zellweger syndrome
(C) Sudden infant death syndrome
(D) von Gierke's disease.

16. The ketone body that does not possess a keto (C=O) group

(A) Acetone (B) Acetoacetate
(C) β-hydroxybutyrate (D) Acetic acid.

17. The ketone body not metabolised

(A) Acetone (B) Acetoacetic acid
(C) β-hydroxy butyrate (D) None of the above.

18. The synthesis of ketone bodies occurs in

(A) Muscle (B) Liver (C) Brain (D) Kidney.

19. The precursor for the synthesis of ketone bodies (ketogenesis)

(A) Acetyl CoA (B) Malonyl CoA
(C) Propionyl CoA (D) Acetate.

20. The ketone bodies can be utilized by the following tissues/organs except

(A) Brain (B) Liver
(C) Skeletal muscle (D) Cardiac muscle.

21. During prolonged starvation, ketone bodies are the major energy sources for

(A) Skeletal muscle (B) cardiac muscle
(C) Brain (D) Kidney.

22. Liver cannot utilize ketone bodies due to lack of the enzyme

(A) Thiolase (B) Thiophorase
(C) HMG CoA lyase
(D) β-hydroxybutyrate dehydrogenase.

23. Ketosis is commonly associated with the following two disorders

(A) Atherosclerosis, von Gierke's disease
(B) Diabetes mellitus, starvation
(C) Fatty liver, atherosclerosis
(D) Nephrotic syndrome, diabetes mellitus.

24. Ketoacidosis is commonly observed in

(A) Diabetes mellitus (B) Starvation
(C) Atherosclerosis (D) Fatty liver.

25. The enzyme machinery for the *de novo* synthesis of fatty acids is located in

(A) Mitochondria (B) Cytosol
(C) Nucleus (D) Peroxisomes.

26. Acetyl CoA from the mitochondria is transported to the cytosol after it is converted to

(A) Malonyl CoA (B) Citrate
(C) Oxaloacetate (D) Malate.

27. One of the following is the regulatory enzymes in fatty acid biosynthesis

(A) Acyl carrier protein
(B) Palmitoyl CoA thioesterase
(C) Acetyl CoA carboxylase
(D) Malonyl CoA-ACP transacylase.

28. **Acetyl CoA carboxylase catalyses the incorporation of CO_2 to produce malonyl CoA. This CO_2 corresponds to the following carbon position when palmitate is synthesized**
(A) 1 (B) 16 (C) 8
(D) None of the above.

29. **Fatty acid synthase is a multienzyme complex involved in fatty acid synthesis. However, one of the following enzymes of fatty acid synthesis is not a component of fatty acid synthase complex**
(A) Acetyl CoA carboxylase
(B) Ketoacyl synthase (C) Enoyl reductase
(D) Acetyl transacylase.

30. **The reducing equivalents for the synthesis of fatty acids are supplied by**
(A) $FADH_2$ (B) NADH
(C) $FMNH_2$ (D) NADPH.

31. **The number of fatty acid(s) that can be simultaneously synthesized by fatty acid synthase complex**
(A) 1 (B) 2 (C) 3 (D) 4.

32. **The synthesis of fatty acids is increased by one of the following compounds by promoting the activity of acetyl CoA carboxylase**
(A) Citrate (B) Palmitoyl CoA
(C) Malonyl CoA (D) Oxaloacetate.

33. **The synthesis of fatty acids is promoted by following hormone**
(A) Glucagon (B) Insulin
(C) Epinephrine (D) Norepinephrine.

34. **The NADPH utilized for the synthesis of fatty acids is obtained from**
(A) Glycolysis (B) Krebs cycle
(C) Hexose monophosphate shunt
(D) β-oxidation

35. **The final product of the fatty acid synthase complex**
(A) Acetyl CoA (B) Acyl CoA
(C) Palmitate (D) Stearate.

36. **Arachidonic acid, derived from one of the following phospholipids, serves as a substrate for prostaglandin synthesis**
(A) Lecithin (B) Cephalin
(C) Phosphatidylserine
(D) Phosphatidylinositol.

37. **Niemann-Pick disease is due to a defect in the enzyme**
(A) Ceramidase (B) Sphingomyelinase
(C) β-galactosidase (D) β-glucosidase.

38. **Faber's disease is due to a defect in the enzyme**
(A) Sphigomyelinase (B) β-glucosidase
(C) Ceramidase (D) β-galactosidase.

39. **The total body content of cholesterol is around**
(A) 50 g. (B) 100 g. (C) 150 g. (D) 200 g.

40. **The carbon atoms of cholesterol in its biosynthesis are derived from**
(A) Acetate of acetyl CoA
(B) CoA of acetyl CoA
(C) NADPH (D) ATP.

41. **β-Hydroxy β-methylglutaryl CoA (HMG CoA) is the key intermediate in the biosynthesis of cholesterol. HMG CoA for cholesterol synthesis is produced in the**
(A) Mitochondria (B) Cytosol
(C) Mitochondria and cytosol
(D) Microsomes.

42. **The rate limiting enzyme in cholesterol biosynthesis**
(A) Thiolase (B) HMG CoA synthase
(C) HMG CoA reductase
(D) Mevalonate kinase.

43. **The reducing equivalents for cholesterol biosynthesis are supplied by**
(A) NADH (B) $FADH_2$
(C) NADPH (D) $FMNH_2$.

44. **One of the following coenzymes may be also synthesized while cholesterol biosynthesis occurs**
(A) Coenzyme A (B) Coenzyme Q
(C) Thiamine pyrophosphate
(D) Flavin mononucleotide.

45. **HMG CoA reductase is more active as**
(A) Phosphorylated enzyme
(B) Diphosphorylated enzyme
(C) Acetylated enzyme
(D) Hydroxylated enzyme.

46. **Conversion of squalene to cholesterol involves the following reaction(s)**
(A) Hydroxylation (B) cyclization
(C) Utilization of O_2 and NADPH
(D) All of the above.

47. **One of the following hormones promotes the synthesis of cholesterol**
(A) Thyroxine (B) Glucagon
(C) Glucocorticoids (D) None of the above.

48. The drugs compactin and lovastatin inhibit cholesterol biosynthesis by acting as

(A) Competitive inhibitors
(B) Non-competitive inhibitors
(C) Allosteric inhibitors
(D) Feedback inhibitors.

49. Cholesterol can serve as a precursor for the synthesis of

(A) Bile acids (B) Steroid hormones
(C) Coprostanol (D) All of the above.

50. The following compound is administered for the treatment of cholelithiasis

(A) Compactin (B) Lovastatin
(C) Chenodiol (D) Lecithin.

51. One of the following intermediates of cholesterol biosynthesis is converted to vitamin D_3 (cholecalciferol) in the skin by UV rays

(A) Squalene (B) Lanosterol
(C) Mevalonate (D) 7-dehydrocholesterol.

52. The ratio of esterified cholesterol to free cholesterol in circulation is around

(A) 1 : 1 (B) 2 : 1 (C) 3 : 1 (D) 1 : 3.

53. The activity of lecithin-cholesterol acyltransferase (LCAT) is associated with

(A) High density lipoproteins
(B) Low density lipoproteins
(C) Chylomicrons
(D) Very low density. lipoproteins.

54. Lecithin-cholesterol acyltransferase catalyses the transfer of fatty acid from the following compound to cholesterol

(A) Phosphatidyl choline
(B) Phosphatidyl ethanolamine
(C) Phosphatidyl inositol
(D) Plasmalogens.

55. In healthy individuals, the total plasma cholesterol concentration is in the range of

(A) 50-100 mg/dl (B) 50-150 mg/dl
(C) 100-150 mg/dl (D) 150-200 mg/dl.

56. Elevation of the plasma cholesterol, associated with one of the following lipoproteins, protects the body from the risk of coronary heart diseases

(A) Low density lipoproteins
(B) Very low density lipoproteins
(C) High density lipoproteins
(D) Chylomicrons.

57. The occurrence of atherosclerosis and coronary heart disease is usually associated with the elevation in plasma

(A) LDL-cholesterol (B) VLDL-cholesterol
(C) HDL-cholesterol (D) All of the above.

58. Hypercholesterolemia is observed in

(A) Hypothyroidism (B) Obstructive jaundice
(C) Nephrotic syndrome
(D) All of the above.

59. Hypercholesterolemia can be controlled by

(A) Dietary polyunsaturated fatty acids
(B) Dietary fiber
(C) Administration of lovastatin
(D) All of the above.

60. One of the following plasma lipoproteins is not usually separated by electrophoresis

(A) Chylomicrons
(B) Low density lipoproteins
(C) Free fatty acids-albumin
(D) High density lipoproteins.

61. The highest quantity of lipid and lowest concentration protein are found in

(A) Chylomicrons
(B) Very low density lipoproteins
(C) Low density lipoproteins
(D) High density lipoproteins.

62. The deficiency of LDL receptors causes the hyperlipoproteinemia

(A) Type I (B) Type IIa
(C) Type IIb (D) Type III.

63. The plasma enzyme lecithin-cholesterol acyl transferase (LCAT) catalyses the esterification of free cholesterol present in the extrahepatic tissues and transfers to

(A) Chylomicrons
(B) Low density lipoproteins
(C) Very low density lipoproteins
(D) High density lipoproteins.

64. The lipoproteins that are synthesized in the intestine that transport triacylglycerols to various tissues

(A) Chylomicrons
(B) Very low density lipoproteins
(C) Low density lipoproteins
(D) High density lipoproteins.

65. Type I hyperlipoproteinemia, due to the deficiency of lipoprotein lipase, is associated with increased plasma concentration of
(A) Chylomicrons and triacylglycerols
(B) Low density lipoproteins and cholesterol
(C) Very low density lipoproteins and cholesterol
(D) Very low density lipoproteins and triacylglycerols.

66. Type IIa hyperlipoproteinemia, due to a defect in LDL receptors, is characterized by increased plasma concentration of
(A) LDL and triacylglycerols
(B) LDL and cholesterol
(C) LDL and chylomicrons
(D) Chylomicrons and cholesterol.

67. Abetalipoproteinemia is characterized by the total absence of
(A) Chylomicrons
(B) Low density lipoproteins
(C) Very low density lipoproteins
(D) High density lipoproteins.

68. In Tangier disease, there is a total absence of
(A) Chylomicrons
(B) Low density lipoproteins
(C) Very low density lipoproteins
(D) High density lipoproteins.

69. The normal concentration of lipids in the liver is around
(A) 5% (B) 10% (C) 15% (D) 20%.

70. Fatty liver may occur due to
(A) Increased synthesis of triacylglycerols
(B) Impairment in lipoprotein synthesis
(C) Deficiency of lipotropic factors
(D) Any one of the above.

71. Fatty liver is primarily due to the accumulation of
(A) Triacylglycerols (B) Phospholipids
(C) Cholesterol (D) Glycolipids.

72. Impairment in the synthesis of lipoproteins by liver is associated with
(A) A defect in phospholipid synthesis
(B) A block in apoprotein formation
(C) A failure in the secretion of lipoproteins
(D) All of the above.

73. The substances, the deficiency of which causes fat accumulation in the liver are referred to as
(A) Lipoprotein factors(B) Lipotropic factors
(C) Phospholipids (D) Vitamin factors.

74. The following substances are lipotropic factors except
(A) Choline (B) Methionine
(C) Betaine (D) Cysteine.

75. The lipotropic factors are largely concerned with a reaction, namely
(A) Transamination (B) Transmethylation
(C) Deamination (D) Transdeamination.

76. The following hormonal related disorder(s) may cause obesity
(A) Cushing's syndrome (B) Hypothyroidism
(C) Insulinoma (D) All of the above.

77. The mitochondria of brown adipose tissue are characterized by the following except
(A) High proprotion of cytochromes
(B) Very active oxidation of NADH
(C) Energy liberated as heat
(D) Adequate synthesis of ATP.

78. Thermogenin is a protein isolated from mitochondria of brown adipose tissue. Its function is comparable to a/an
(A) Lipotropic factor (B) Uncoupler
(C) ETC inhibitor (D) Primer.

79. Atherosclerosis is characterized by thickening of arteries due to the accumulation of
(A) Cholesterol and cholesterol ester
(B) Fibrous tissue and collagen
(C) Proteoglycans and calcium
(D) All of the above.

80. The development of atherosclerosis and the risk for coronary heart disease is usually correlated with plasma concentration of
(A) High density lipoproteins
(B) Low density lipoproteins
(C) Chylomicrons
(D) Very low density lipoproteins.

81. The following disorder(s) is (are) associated with atherosclerosis
(A) Diabetes mellitus (B) Hypothyroidism
(C) Nephrotic syndrome
(D) All of the above.

82. The low incidence of cardiovascular disorders is associated with increased plasma concentration of
(A) Chylomicrons
(B) Low density lipoproteins
(C) Very low density lipoproteins
(D) High density lipoproteins.

83. Alcohol (over)consumption for a long period leads to

(A) Decreased fatty acid oxidation

(B) Impairment in citric acid cycle

(C) Increased triacylglycerol synthesis and fatty liver

(D) All of the above.

84. A patient of hypercholesterolemia was prescribed statin [atorvastatin] to inhibit the enzyme HMG CoA reductase and reduce cholesterol biosynthesis. Statins should be taken at the following specified timing for optical activity.

(A) 5 AM (B) 10 AM (C) 5 PM (D) 10 PM

85. A 50 year old man had elevated total serum cholesterol (250 mg/dl) and increased lipoprotein-a (Lp-a) –90 mg/dl (normal <30 mg/dl)] This man is at risk for one of the following diseases.

(A) Fatty liver (B) Coronary heart disease

(C) Diabetes mellitus

(D) Nephrotic syndrome

Answers

1. (**B**) 2. (**D**) 3. (**C**) 4. (**D**) 5. (**B**) 6. (**C**) 7. (**B**) 8. (**B**) 9. (**A**) 10. (**C**) 11. (**D**) 12. (**B**) 13. (**C**) 14. (**A**) 15. (**B**) 16. (**C**) 17. (**A**) 18. (**B**) 19. (**A**) 20. (**B**) 21. (**C**) 22. (**B**) 23. (**B**) 24. (**A**) 25. (**B**) 26. (**B**) 27. (**C**) 28. (**D**) 29. (**A**) 30. (**D**) 31. (**B**) 32. (**A**) 33. (**B**) 34. (**C**) 35. (**C**) 36. (**D**) 37. (**B**) 38. (**C**) 39. (**C**) 40. (**A**) 41. (**B**) 42. (**C**) 43. (**C**) 44. (**B**) 45. (**B**) 46. (**D**) 47. (**A**) 48. (**A**) 49. (**D**) 50. (**C**) 51. (**D**) 52.(**C**) 53. (**A**) 54. (**A**) 55. (**D**) 56. (**C**) 57. (**A**) 58. (**D**) 59. (**D**) 60. (**C**) 61.(**A**) 62. (**B**) 63. (**D**) 64. (**A**) 65. (**A**) 66.(**B**) 67. (**B**) 68. (**D**) 69. (**A**) 70. (**D**) 71. (**A**) 72. (**D**) 73. (**B**) 74. (**D**) 75. (**B**) 76. (**D**) 77. (**D**) 78. (**B**) 79. (**D**) 80. (**B**) 81. (**D**) 82. (**D**) 83. (**D**) 84. (**D**) 85. (**B**).

Clinical Case Study 14.1 (Source : Biochemistry, Page 333)

A 30-year old man complained chest pain. His clinical and biochemical investigations (ECG changes, isoenzymes of creatine phosphokinase etc.) indicated that he suffered a mild myocardial infarction. Timely intervention and appropriate treatment saved him from death. The lipid profile data of this man are given here.

Parameter	*Subject (mg/dl)*	*Reference range*
Serum total cholesterol	450	150-200
LDL-cholesterol	380	<100
HDL-cholesterol	40	40-60
VLDL-cholesterol	30	20-40
Serum triglycerides	150	75-150

1. **What are the biochemical alterations and what is the probable diagnosis?**
2. **Which lipoprotein receptors are decreased in this disorder?**
3. **What is the suggested mode of treatment?**

Ans. :

1. This man had highly elevated total and LDL-cholesterol while other lipid profile parameters are within the normal limits. This is a case of **familial hypercholesterolemia (hyperlipoproteinemia type IIa)**.
2. The victims of this disorder have decreased number of LDL receptors.
3. Lipid-lowering drugs, besides low cholesterol, low fat diet and regular exercise are useful for these people.

Clinical Case Study 14.2 (Source : Biochemistry, Page 324)

The laboratory data on the lipid profile of a 35-year old man are given here.

Parameter	*Subject (mg/dl)*	*Reference range*
Serum total cholesterol	350	150-200
LDL-cholesterol	225	<100
HDL-cholesterol	35	40-60
VLDL-cholesterol	90	20-40
Serum triglycerides	450	75-150

Further, the electrophoresis of plasma lipoproteins showed a thick band of intermediate density lipoproteins (IDL).

1. **Discuss lipid profile changes along with the disorder.**
2. **What is the probable cause of this disorder?**
3. **What are the associated complications?**

Ans. :

1. The serum triglycerides and LDL-cholesterol were elevated in this subject, besides a distinct band of IDL on electrophoresis. These findings suggest that this is a case of **type III hyperlipoproteinemia**.
2. Although the etiopathology of this disorder is not known, it is attributed to an abnormality in apoprotein E.
3. The sufferers of this disease are at a high risk for atherosclerosis and coronary heart diseases.

Clinical Case Study 14.3 (Source : Biochemistry, Page 324)

A 55-year old diabetic man had been on treatment with oral antidiabetic drugs for over a decade. Despite the physician's advice, he could not give up his drinking habit. The lipid profile data

Parameter	*Subject (mg/dl)*	*Reference range*
Serum total cholesterol	250	150-200
LDL-cholesterol	110	<100
HDL-cholesterol	40	40-60
VLDL-cholesterol	100	20-40
Serum triglycerides	500	75-150

Further, the serum lipoprotein electrophoretic pattern indicated an increase in the concentration of chylomicrons and VLDL.

1. **What are the biochemical changes and what is the disorder?**
2. **Which lipid parameter is most excessively produced?**
3. **What is the nutritional advice to these patients.**

Ans. :

1. The diabetic patient and an alcoholic had elevated levels of triglycerides and VLDL-cholesterol. This is a case of **type V hyperlipoproteinemia**.
2. Triglycerides.
3. The sufferers of this disease are advised low calorie and low fat diet.

Clinical Case Study 14.4
(Source : Biochemistry, Page 287)

A 6 month old male infant came to OPD with complaints of abdominal distension, developmental delay, and failure to these on examination infant has hepatospleenomegaly, Blood investigations showed increased cholesterol and triglycerides. Ultrasound indicated bilateral adrenal calcification and organomegaly.

1. **What is the diagnosis?**
2. **What is the enzyme defect in this disease?**
3. **What is the mechanism for developing symptoms?**

Ans. :

1. **Wolman disease** which is an autosomal recessive lysosomal storage disorder
2. Lysosomal acid lipase
3. Lysosomal acid lipase is essential for metabolism of triglycerides and cholesterol esters. Its defeciency leads to elevated level of TG and cholesterol, and cause their deposition in excessive amount in the visceral organs resulting in organomegaly and abdominal distension, diarrhea and failure to thrive.

Clinical Case Study 14.5
(Source : Biochemistry, Pages 317-318)

A 42 year old female was found to have high serum cholesterol and LDL cholesterol. She had family history of father suffering from coronary artery disease. Physician advised medication to decrease cholesterol levels.

1. **What is the diagnosis?**
2. **What are the normal levels of serum cholesterol and TG?**
3. **What are the products derived from cholesterol?**
4. **What is the treatment?**

Ans. :

1. **Hypercholesterolemia.**
2. Normal levels of serum cholesterol-150-200 mg/dl and TG–75-150 mg/dl.
3. Bile acids, steroid hormones, adrenocortical hormones, and vitamin D.
4. Consumption of polyunsaturated fatty acids (PUFA), change in lifestyle and use of statins (lovastatin, simvastatin etc) which competitively inhibit HMG CoA reductase and decrease cholesterol synthesis.

Clinical Case Study 14.6
(Source : Biochemistry, Pages 324-326)

A 30 years old obese male who was known diabetic complains of indigestion, generalized weakness and abdominal discomfort. Laboratory reports revealed elevated triglycerides (dyslipidemia) on ultra sound fibrotic changes in liver, and on microscopy of liver tissue infiltration, ballooning of cells

1. **What is the diagnosis?**
2. **What is the cause of this disorder?**
3. **What is the biochemical mechanism behind this disease?**
4. **What are factors which prevent this disease?**

Ans. :

1. **Fatty liver.**
2. It is occurs due to two main reasons – increased triacylglycerol synthesis and/or impaired synthesis of lipoproteins.
3. Refer Chapter 14, pages 324-326.
4. Lipotropic factors – the substances that are necessary for normal mobilization of fat from liver to prevent fatty liver. The deficiency of these factors leads to fatty liver. e.g., choline, betaine, methionine, inositol, folic acid vitamin B_{12}.

For more clinical case studies, refer 7.7, 16.1, 19.1, 21.1, 21.2

VIVA/Short Questions

1. Why should fats be the fuel reserve of the body?

They are highly concentrated form of energy (9 Cal/g) and can be stored in an anhydrous form (no association with water).

2. What do you understand by dynamic state of body lipids?

Lipids are continuously being degraded and resynthesized and thus are in a dynamic state.

3. How is fat mobilized form adipose tissue?

Hormone-sensitive triacylglycerol lipase removes the fatty acids either from carbon 1 or 3 of TG to form diacylglycerol which in turn is cleaved by additional lipases. The complete degradation of TG to glycerol and free fatty acids is known as lipolysis.

4. What is the fate of glycerol?

Glycerol is transported to liver where it is activated to glycerol 3-phosphate which may be used for the synthesis of triacylglycerols, phospholipids, or enter glycolysis.

5. What is β-oxidation?

The oxidation of the fatty acids on the β-carbon atom, with a sequential removal of a two carbon fragment, acetyl CoA.

6. Name some tissues which cannot oxidize fatty acids for energy?

Brain, erythrocytes and adrenal medulla.

7. What is the role of carnitine in fat metabolism?

Carnitine (β-hydroxy γ-trimethyl-aminobutyrate) is the carrier of activated fatty acids from cytosol to the mitochondria for their oxidation.

8. What is sudden infant death syndrome?

An unexpected death of healthy infants, usually overnight. At least 10% of SIDS is due to the deficiency of the enzyme medium chain acyl CoA dehydrogenase.

9. What do you know about Refsum's disease?

Refsum's disease is due to a defect in the α-oxidation of fatty acids. The patients accumulate large quantities of phytanic acid which cannot undergo β-oxidation due to the presence of a methyl group on carbon-3.

10. What are ketone bodies?

Acetone, acetoacetate and β-hydroxybutyrate.

11. What do you know about ketogenesis?

The synthesis of ketone bodies (ketogenesis) occurs in the liver. The enzymes of ketogenesis are located in the mitochondrial matrix.

12. How are ketone bodies utilized?

Acetoacetate and β-hydroxybutyrate serve as important sources of energy for peripheral tissues such as skeletal muscle, cardiac muscle, renal cortex etc.

13. What do you understand by the terms ketonemia, ketonuria and ketosis?

Ketonemia refers to an increase in the concentration of ketone bodies in the circulation while ketonuria represents their excretion into urine. The overall picture of ketonemia and ketonuria is referred to as ketosis.

14. What is ketoacidosis?

The ketone bodies — acetoacetate and β-hydroxybutyrate are strong acids. Increase in their concentration in the blood would cause acidosis.

15. What do you understand by *de novo* synthesis of fatty acids?

De novo (a new) synthesis of fatty acids occurs when the dietary consumption of carbohydrates and amino acids is in excess. The enzyme machinery for fatty acid synthesis is localized in the cytosomal fraction.

16. What is the source of carbon atoms in fatty acid synthesis and how is it made available?

Acetyl CoA is the ultimate source of all the carbon atoms in fatty acids. It is produced in mitochondria by the oxidation of pyruvate and fatty acids, degradation of carbon skeleton of amino acids, and from ketone bodies.

17. What is the first reaction in fatty acid synthesis?

Acetyl CoA is carboxylated to malonyl CoA by a biotin dependent enzyme acetyl CoA carboxylase. This is a regulatory enzyme in fatty acid biosynthesis.

18. What do you know about fatty acid synthase (FAS) complex?

The fatty acid synthesis occurs through the participation of a multifunctional enzyme known (FAS) complex. It is a dimer with two identical units. Each monomer possesses the activities of seven different enzymes and an acyl carrier protein (ACP).

19. How is fatty acid synthesis regulated?

By the enzyme acetyl CoA carboxylase, hormones (insulin promotes while glucagon inhibits) and dietary manipulation (fasting or high fat diet decrease).

20. How does triacylglycerol synthesis occur?

The fatty acids and glycerol are activated and then in a series of reactions get converted to triacylglycerol (fat).

21. What do you know about Niemann-Pick disease?

An inherited disorder due to a defect in the enzyme sphigomyelinase. This causes the accumulation of sphingomyelins in liver and spleen resulting in enlargement of these organs.

22. What is Faber's disease?

A defect in the enzyme ceramidase, characterized by skeletal deformation, subcutaneous nodules, dermatitis and mental retardation.

23. Mention salient features of cholesterol biosynthesis?

Acetyl CoA provides all the carbon atoms in cholesterol. The reducing equivalents are supplied by NADPH while ATP provides energy. The important intermediates of cholesterol biosynthesis include HMG CoA, mevalonate, isoprenoid units and squalene. Cholesterol biosynthesis is regulated by a feedback control of the enzyme HMG CoA reductase.

24. How is cholesterol biosynthesis regulated?

Mainly controlled by the rate limiting enzyme, HMG CoA reductase. By feedback mechanism (cholesterol inhibits its own synthesis), hormonal control (glucagon decreases while insulin increases cholesterol synthesis).

25. How is cholesterol degraded?

Cholesterol is converted to bile acids (glycocholic acid, taurocholic acid), serves as a precursor for the synthesis of steroid hormones (progesterone, cortisol, estradiol), vitamin D.

26. What do you know about bile salts?

The conjugated bile acids (glycocholic acid, taurocholic acid) exist as sodium and potassium salts which are known as bile salts. Bile salts serve as emulsifying agents and facilitate the digestion and absorption of fats and fat soluble vitamins.

27. What is cholelithiasis?

Cholelithiasis (cholesterol gall stone disease) is the deposition of cholesterol crystals in the gall bladder.

28. How does cholesterol serve as a precursor for the synthesis of vitamin D.

7-Dehydrocholesterol, an intermediate in the synthesis of cholesterol, is converted to cholecalciferol by ultraviolet rays in the skin.

29. How is cholesterol transported in the circulation?

In the plasma lipoproteins in two forms. About 70-75% in an esterified form with long chain fatty acids and about 25-30% as free cholesterol.

30. What is LCAT and what is its importance?

Lecithin-cholesterol acyltransferase (LCAT) is a plasma enzyme. It catalyses the transfer of a fatty acid from lecithin to cholesterol to produce cholesterol ester. In this manner, cholesterol from the peripheral tissues is transported to the liver for degradation and excretion.

31. What do you know about hyper-cholesterolemia?

Increase in plasma cholesterol (>200 mg/dl) concentration, observed in several disorders such as diabetes mellitus, hypothyroidism, obstructive jaundice and nephrotic syndrome.

32. What measures do you suggest to control hypercholesterolemia?

Consumption of a diet rich in polyunsaturated fatty acids (PUFA) and fiber, and low in carbohydrates, besides the administration of drugs (lovastatin, clofibrate).

33. What do you know about lipoproteins?

Lipoproteins are molecular complexes that consist of lipids and proteins. They fuction as transport vehicles for lipids in blood plasma. There are five major classes of lipoproteins in human plasma namely, chylomicrons, very low density lipoproteins (VLDL), low density lipoproteins (LDL), high density lipoproteins (HDl) and free fatty acids-albumin.

34. How are plasma lipoproteins separated?

By electrophoresis into chylomicrons, low density lipoproteins, very low density lipoproteins and high density lipoproteins.

35. What are apolipoproteins?

Protein components of lipoproteins and are closely involved in the metabolism of the later, e.g., AI, AIII, B_{100}, C_1, CII.

36. What is the role of lipoprotein lipase?

Hydrolyses a portion of the triacylglycerols present in chylomicrons and VLDL to liberate free fatty acids and glycerol.

37. What are the functions of high density lipoproteins?

Transport cholesterol (as ester) from peripheral tissues to liver for its degradation and excretion.

38. What do you know about hyperlipo-proteinemias?

Elevation in one or more of the lipoproteins fractions constitutes hyperlipoproteinemias (or hyperlipidemias). Fredericksons classification of hyperlipoproteinemias (Types-I, IIa, IIb, III, IV and V) based on the electrophoretic pattern of plasma lipoproteins, is used to understand these disorders.

39. What is abetalipoproteinemia?

Abetalipoproteinemia is a rare disorder due to a defect in the synthesis of apoprotein B. It is characterized by a total absence of LDL in plasma.

40. What is Tangier disease?

Familial α-lipoprotein deficiency, characterized by a total absence of HDL, adversely affecting the reverse transport of cholesterol.

41. What do you understand by fatty liver?

The excessive accumulation of triacylglycerols in the liver (normal <5% by weight).

42. What are the causative factors for impairment in lipoprotein synthesis that result in fatty liver?

Impairment in phospholipid synthesis is associated with the dietary deficiency of lipotropic facotrs (choline, betaine, inositol).

43. What do you know about lipotropic factors?

Lipotropic factors are substances the deficiency of which causes fatty liver. e.g., choline, betaine, methionine and inositol. They are directly or indirectly concerned with the synthesis of phospholipids.

44. What is obesity?

Obesity is an abnormal increase in the body weight due to excessive fat deposition, more than 20% and 25% of body weight in men and women respectively.

45. What is anorexia nervosa?

Characterized by a total loss of appetite. This is more a psychiatric disease, mostly seen in females in the age group 10-30 years.

46. What do you know about atherosclerosis?

A complex disease characterized by thickening of arteries due to the accumulation of lipids (particularly cholesterol), collagen, fibrous tissue, calcium deposits etc.

47. What is the relation between high density lipoproteins (HDL) and coronary heart disease?

The increased levels of plasma HDL are correlated with low incidence of cardiovascular disorders. Women have higher plasma HDL and are less prone to heart diseases. This is attributed to estrogens in women.

48. What do you know about lipoprotein a (Lp-a)?

Lp-a is almost identical in structure to LDL, and it inhibits fibrinolysis. Recent studies suggest that elevation in plasma Lp-a (>30 mg/dl) is associated with increased risk of coronary heart disease.

15. Metabolism of Amino Acids

MCQs

1. The following two amino acids constitute about 50% of the amino acid pool of the body

(A) Glycine and alanine
(B) Glutamate and glutamine
(C) Aspartate and asparagine
(D) tyrosine and phenylalanine.

2. Protein degradation is facilitated by a polypeptide namely

(A) Thrombin (B) Plasmin
(C) Ubiquitin (D) Ubiquinone.

3. The following is (are) the source (s) of the amino acid pool in the body

(A) Protein turnover (B) Dietary protein
(C) Synthesis of non-essential amino acids
(D) All of the above.

4. Normally, the contribution of energy from the amino acid degradation to the body is in the range of

(A) 10-15% (B) 25-30%
(C) 40-45% (D) 55-60%.

5. Transamination is dependent on the coenzyme

(A) Thiamine pyrophosphate
(B) Flavin adenine dinucleotide
(C) Biotin (D) Pyridoxal phosphate.

6. Transamination is characterized by the following except it is

(A) Reversible
(B) Involved in the liberation of free NH_3
(C) Involved in the synthesis of non-essential amino acids
(D) Dependent of the coenzyme pyridoxal phosphate.

7. One of the following amino acids cannot undergo transamination

(A) Glycine (B) Alanine
(C) Lysine (D) Glutamate.

8. The δ-amino group of the following amino acid can be transaminated

(A) Lysine (B) Arginine
(C) Ornithine (D) Histidine.

9. The amino acid predominantly involved in transdeamination reaction

(A) Alanine (B) Glutamate
(C) Aspartate (D) Glutamine.

10. The amino acid that serves as the collection centre for the amino groups in transamination

(A) Glutamate (B) Glutamine
(C) Aspartate (D) Glycine.

11. L-Glutamate dehydrogenase is dependent on the coenzyme
(A) NAD^+ (B) $NADP^+$
(C) NAD^+ or $NADP^+$ (D) FAD.

12. The following statements about glutamate dehydrogenase (GDH) are true except it
(A) Is a mitochondrial enzyme
(B) Contains zinc
(C) Is an allosteric enzyme
(D) Requires FAD.

13. Oxidative deamination results in the release of free ammonia. The enzyme predominantly associated with this process is
(A) L-amino acid oxidase
(B) D-amino acid oxidase
(C) Glutamate dehydrogenase
(D) Alanine transaminase.

14. L-Amino acid oxidase is dependent on the coenzyme
(A) FMN (B) FAD (C) NAD^+ (D) $NADP^+$.

15. The following amino acid(s) can undergo non-oxidative deamination
(A) Serine (B) Cysteine
(C) Histidine (D) All of the above.

16. The source(s) of ammonia formation is (are)
(A) Amino acids
(B) Purines and pyrimidines
(C) Action of intestinal bacteria on urea
(D) All of the above.

17. The transport of ammonia occurs in the form of
(A) Ammonium ions (B) Glutamine
(C) Alanine
(D) Glutamine or alanine.

18. The amino acid that is present at the highest concentration in the blood, serving as storage and transport form of ammonia
(A) Alanine (B) Glutamate
(C) Glutamine (D) None of the above.

19. Ammonia is involved (directly or indirectly) in the synthesis of
(A) Non-essential amino acids
(B) Purines and pyrimidines
(C) Acid-base balance (D) All of the above.

20. Reptiles and birds dispose of ammonia from the body as
(A) Ammonium ions (B) Urea
(C) Uric acid (D) Amino acids.

21. Elevation in blood ammonia concentration is most harmful to
(A) Liver (B) Kidney (C) Muscle (D) Brain.

22. During ammonia toxicity, one of the following metabolites of citric acid cycle is depleted in the brain
(A) Citrate (B) Oxaloacetate
(C) α-ketoglutarate (D) Succinate.

23. Toxic effects of ammonia on the brain are mainly due to impairment in
(A) ATP production
(B) Glutamate formation
(C) Glutamine synthesis
(D) None of the above.

24. The excretory end product of protein (amino acid) metabolism in humans
(A) Ammonia (B) Urea
(C) Uric acid (D) Allantoin.

25. Of the two amino groups present in urea, one is derived from ammonia while the other one comes from
(A) Carbamoyl phosphate
(B) Citrulline (C) Aspartate
(D) Glutamate.

26. Carbamoyl phosphate synthase I is characterized by all of the following except it is
(A) A mitochondrial enzyme
(B) Activated by N-acetylglutamate
(C) Involved in the urea synthesis
(D) Involved in catalysing a reversible reaction.

27. One of the following amino acids is continuously regenerated and recycled in urea synthesis
(A) Aspartate (B) Citrulline
(C) Arginine (D) Ornithine.

28. The amino acid that is produced in the mitochondria and transported to cytosol for urea synthesis
(A) Ornithine (B) Citrulline
(C) Arginine (D) Arginosuccinate.

29. One of the following enzymes of urea cycle is exclusively found in liver.
(A) Ornithine transcarbamoylase
(B) Arginosuccinate synthase
(C) Arginosuccinase
(D) Arginase.

30. The number of ATP consumed in urea synthesis
(A) 1 (B) 2 (C) 3 (D) 4.

31. The non-protein amino acid(s) found in urea cycle
(A) Ornithine (B) Citrulline
(C) Arginosuccinate (D) All of the above.

32. The non-essential amino acid synthesized in urea cycle
(A) Arginine (B) Aspartate
(C) Glutamate (D) Ornithine.

33. The rate limiting reaction in urea cycle is catalyzed by the enzyme
(A) Carbamoyl phosphate synthase I
(B) Ornithine transcarbamoylase
(C) Arginosuccinase (D) Arginase.

34. Carbamoyl phosphate synthase I coordinates with one of the following enzymes that liberates ammonia entering urea cycle
(A) Glutaminase
(B) Glutamine synthetase
(C) Glutamate dehydrogenase
(D) Arginase.

35. One of the following metabolites liberated in the urea cycle enters the citric acid cycle
(A) Citrate (B) α-ketoglutarate
(C) Succinate (D) Fumarate.

36. One of the following metabolites of citric acid cycle on transamination produces an amino acid that participates in urea cycle
(A) α-ketoglutarate (B) Oxaloacetate
(C) Succinate (D) Malate.

37. Blood urea estimation is widely used as a screening test for the evaluation of the function of
(A) Liver (B) Heart
(C) Kidney (D) Skeletal muscle.

38. The normal blood urea concentration
(A) 10-20 mg/dl (B) 10-40 mg/dl
(C) 30-60 mg/dl (D) 60-100 mg/dl.

39. Blood urea concentration is elevated in
(A) Acute glomerulonephritis
(B) Polycystic kidney (C) Diabetic coma
(D) All of the above.

40. One of the following is the most abundant non-protein nitrogenous substance in the blood
(A) Urea (B) Uric acid
(C) Creatinine (D) Amino acids.

41. If the blood urea concentration is 100 mg/dl, the blood urea nitrogen (BUN) approximately corresponds to
(A) 25 mg/dl (B) 50 mg/dl
(C) 75 mg/dl (D) 100 mg/dl.

42. The relation between blood urea nitrogen (BUN) and non-protein nitrogen (NPN) is represented by
(A) BUN = $^1/_2$NPN (B) BUN = 2NPN
(C) BUN = NPN (D) BUN = $^1/_4$NPN.

43. The following statements on glycine metabolism are true except it is
(A) Involved in the synthesis of heme
(B) Ketogenic
(C) Capable of donating one carbon unit
(D) Glycogenic.

44. The following protein contains very high content (about 30%) of glycine
(A) Albumin (B) Fibrinogen
(C) Collagen (D) Immunoglobulin G.

45. Glycine contributes to the synthesis of purines at positions
(A) 1, 2 and 3 (B) 4, 5 and 6
(C) 4, 5 and 7 (D) 5, 6 and 8.

46. Glycine combines with the following two amino acids to produce glutathione
(A) Aspartate and cysteine
(B) Glutamate and cysteine
(C) Glutamate and serine
(D) Cysteine and methionine.

47. Benzoic acid, a food preservative, gets detoxified on conjugation with glycine to form
(A) Hippuric acid (B) Glycocholic acid
(C) Glycocolate (D) Oxalic acid.

48. Glycine condenses with one of the following metabolites of citric acid cycle to form δ-amino levulinate which serves as a precursor for heme synthesis
(A) α-ketoglutarate (B) Succinyl CoA
(C) Fumarate (D) Malate.

49. The three amino acids involved in the synthesis of creatine phosphate
(A) Glycine, histidine and methionine
(B) Glycine, arginine and methionine
(C) Glycine, methionine and valine
(D) Leucine, glycine and methionine.

50. Serum creatinine estimation is a valuable index for the assessment of the functioning of
(A) Liver (B) Heart
(C) Kidney (D) Skeletal muscle.

51. Primary hyperoxaluria is characterized by a defect in the enzyme
(A) Glycine synthase
(B) Glycine transaminase
(C) Glycine oxidase (D) Serine dehydratase.

52. Elevation in the urinary oxalate excretion sometimes gets corrected by the supplementation of
(A) Vitamin C (B) Vitamin B_6
(C) Vitamin C (D) Vitamin B_{12}.

53. The metabolic functions of phenylalanine mostly occur after its conversion to
(A) Alanine (B) Tyrosine
(C) Tryptophan (D) Histidine.

54. Tyrosine is characterized by the following except it is
(A) A precursor for serotonin
(B) Involved in the synthesis of catecholamines
(C) Both glucogenic and ketogenic
(D) A precursor for melanin.

55. Phenylalanine hydroxylase converts phenylalanine to tyrosine. This enzyme is dependent on the coenzyme
(A) Tetrahydrofolate
(B) Tetrahydrobiopterin
(C) Dihydrobiopterin (D) Coenzyme Q.

56. The following intermediate of citric acid cycle is produced from tyrosine
(A) Fumarate (B) Succinate
(C) α-ketoglutarate (D) Malate.

57. The black pigment of the skin is due to
(A) Melatonin (B) Serotonin
(C) Melanosomes (D) Melanin.

58. The enzyme responsible for the synthesis of melanin
(A) Tyrosine hydroxylase (B) Tyrosinase
(C) Tyrosine transaminase
(D) Homogentisate oxidase.

59. Thyroid hormones (thyroxine and triiodothyronine) are synthesized from tyrosine residue of the protein
(A) Thyroalbumin (B) Thyroglobulin
(C) Collagen (D) Albumin.

60. The following are the catecholamines except
(A) Dopamine (B) Epinephrine
(C) Serotonin (D) Norepinephrine.

61. Conversion of norepinephrine to epinephrine involves
(A) Carboxylation (B) Methylation
(C) Decarboxylation (D) Hydroxylation.

62. Parkinson's disease is linked with an impairment in the synthesis of
(A) γ-aminobutyrate (B) Serotonin
(C) Noropinephrine (D) Dopamine.

63. Phenylketonuria is due to a defect in the enzyme
(A) Tyrosine hydroxylase
(B) Phenylalanine hydroxylase
(C) Tyrosinase (D) Homogentisate oxidase.

64. The name phenylketonuria (PKU) is coined due to the fact that one of the following metabolites excreted in urine is a keto acid
(A) Phenylpyruvate (B) Phenylacetate
(C) Phenyllactate (D) Phenylalanine.

65. Mental retardation frequently observed in phenylketonuria patients is believed to be due to
(A) Impairment in the transfer and metabolism of aromatic amino acids in the brain
(B) Insufficient synthesis of serotonin
(C) Defect in myelin synthesis
(D) All of the above.

66. The patients of phenylketonuria should be provided a diet
(A) Rich in phenylalanine and tyrosine
(B) Low in phenylalanine and high in tyrosine
(C) Low in phenylalanine and tyrosine
(D) Lacking phenylalanine and tyrosine.

67. Tyrosinemia type II (Richner-Hanhart syndrome) is due to a defect in the enzyme
(A) Phenylalanine hydroxylase
(B) Tyrosine transaminase
(C) Homogentisate oxidase (D) Tyrosinase.

68. The defect in the enzyme homogentisate oxidase causes
(A) Phenylketonuria (B) Tyrosinemia type II
(C) Alkaptonuria (D) Albinism.

69. The urine of the patients of alkaptonuria gives a positive Benedicts test due to the presence of
(A) Glucose (B) Fructose
(C) Phenylpyruvate (D) Homogentisate

70. The metabolic disorder albinism is characterized by the following except
(A) Lack of melanin synthesis
(B) Increased susceptibility to skin cancers
(C) Defect in the enzyme tyrosine hydroxylase
(D) Intolerance to light.

71. Tryptophan performs the following metabolic functions except it
(A) Is both glycogenic and ketogenic
(B) Is involved in the synthesis of catecholamines
(C) Participates in the formation of serotonin
(D) Is the precursor for the synthesis of NAD^+ and $NADP^+$.

72. The enzyme that cleaves the five membered ring of the indole nucleus in tryptophan metabolism

(A) Tryptophan pyrrolase (B) Kynureninase
(C) Tryptophan hydroxylase
(D) Formamidase.

73. One of the following enzymes of tryptophan metabolism is dependent on pyridoxal phosphate

(A) Tryptophan hydroxylase
(B) Tryptophan pyrrolase
(C) Kynureninase
(D) Picolinate carboxylase.

74. Cat cannot synthesize adequate quantities of NAD^+ and $NADP^+$ from tryptophan due to a high activity of the enzyme

(A) Quinolinate phosphoribosyl transferase
(B) Nicotinate phosphoribosyl transferase
(C) Picolinate carboxylase
(D) Kynureninase.

75. Tryptophan in its metabolism releases one of the following amino acids which is glycogenic

(A) Glycine (B) Alanine
(C) Kynurenine (D) Serine.

76. The antituberculous drug isoniazid blocks one of the following enzymes in tryptophan metabolism

(A) Tryptophan pyrrolase
(B) Formamidase (C) Kynureninase
(D) Picolinate carboxylase.

77. The coenzymes required for serotonin (5-hydroxytryptamine) biosynthesis from tryptophan

(A) Tetrahydrobiopterin and pyridoxal phosphate
(B) Tetrahydrobiopterin and biotin
(C) Pyridoxal phosphate and thiamine pyrophosphate
(D) Pyridoxal phosphate and tetrahydrofolate.

78. Serotonin performes the following functions except it is

(A) An inhibitory neurotransmitter in brain
(B) A powerful vasoconstrictor
(C) Involved in the control of blood pressure and body temperature
(D) Necessary for the motility of gastrointestinal tract.

79. Iproniazid inhibits the following enzyme and elevates serotonin levels

(A) Tryptophan hydroxylase
(B) Aromatic amino acid decarboxylase
(C) Tryptophan decarboxylase
(D) Monoamine oxidase.

80. In mammals, the largest amount of serotonin is synthesized by the

(A) Liver (B) Brain (C) Intestine (D) Heart.

81. Malignant carcinoid syndrome is characterized by

(A) Increased serotonin production by argentaffin cells
(B) Impairment in the synthesis of NAD^+ and $NADP^+$
(C) Elevated urinary excretion of 5-hydroxyindole acetate
(D) All of the above.

82. The chemical name of melatonin

(A) 5-hydroxytryptophan
(B) 5-hydroxytryptamine
(C) N-acetylserotonin
(D) N-acetyl 5-methoxyserotonin.

83. The compound that is closely associated with the circadian rhythms or diurnal variations

(A) Serotonin (B) Dopamine
(C) Melatonin (D) Norepipephrine.

84. Hartnup's disease is due to an impairment in the absorption and transport of tryptophan and other neutral amino acids. The name Hartnup refers to

(A) The scientist who described this disorder first
(B) The family name of the first patient
(C) The name of the place where the patient lived
(D) None of the above.

85. The predominant function of the following sulfur-containing amino acid is its occurrence in protein structure, since it has no other metabolic function

(A) Methionine (B) Cysteine
(C) Homocysteine (D) Cystine.

86. One of the following sulfur containing amino acids is required for the initiation of protein biosynthesis

(A) Methionine (B) Cysteine
(C) Homocysteine (D) Cystine.

87. Transmethylation involves the transfer of methyl group from

(A) Methionine (B) S-adenosylmethionine
(C) S-adenosylhomocysteine
(D) S-adenosylcysteine.

88. The number of high energy phosphates consumed in synthesis of S-adenosylmethionine

(A) 1 (B) 2 (C) 3 (D) 4.

89. The following statements on transmethylation are true except

(A) S-adenosylmethionine donates methyl group
(B) S-adenosylhomocysteine is formed during transmethylation
(C) One ATP is utilized in this process
(D) The enzymes-methyltransferases participate.

90. The formation of one of the following compounds does not involve transmethylation reaction, despite the participation of S-adenosylmethionine

(A) Creatine (B) Melatonin
(C) Choline (D) Spermine.

91. Methionine donates the following group/atom for the synthesis of cysteine

(A) Carboxyl and amino groups
(B) Sulfur only
(C) Sulfur and carboxyl group
(D) Sulfur and amino groups.

92. Transmethylation involves the formation of the following products except

(A) Creatine (B) Choline
(C) Melatonin (D) Norepinephrine.

93. Cysteine participtes in the synthesis of

(A) Glutathione and taurine
(B) Coenzyme A and active sulfate
(C) Glucose and fat (D) All of the above.

94. The inborn error(s) of sulfur containing amino acid metabolism associated with the development of kidney stones

(A) Cystinuria (B) Cystinosis
(C) Homocystinuria I (D) All of the above.

95. Homocystinuria type I is due to a defect in the enzyme

(A) Cystine reductase
(B) Cystathionine synthase
(C) N^5-N^{10}-methylene THF reductase
(D) Cysteine dioxygenase.

96. The vitamin involved in one carbon metabolism, in addition to folic acid

(A) Biotin (B) Thiamine
(C) Vitamin B_6 (D) Vitamin B_{12}.

97. The amino acid from which the formimino (–CH–NH) group is derived for participation in one carbon metabolism

(A) Glycine (B) Histidine
(C) Serine (D) Tryptophan.

98. The prominent non-amino acid donor of one carbon fragments

(A) Adenine (B) Thymidylate
(C) Choline (D) Creatine.

99. The compound(s) involved in one carbon metabolism with reference to methyl group

(A) S-adenosylmethionine
(B) Tetrahydrofolate (C) Vitamin B_{12}
(D) All of the above.

100. The branched chain amino acid that is both glycogenic and ketogenic

(A) Valine (B) Leucine
(C) Isoleucine (D) None of the above.

101. The branched chain amino acid that is only ketogenic

(A) Valine (B) Leucine
(C) Isoleucine (D) None of the above.

102. The metabolic defect associated with branched chain amino acid metabolism

(A) Alkaptonuria (B) Hartnup's disease
(C) Maple syrup urine disease
(D) Albinism.

103. The maple syrup urine disease is characterized by

(A) A defect in the enzyme α-keto acid dehydrogenase
(B) Elevated urinary excretion of branched chain amino acids and their keto acids
(C) Acidosis, lethargy, convulsions and mental retardation
(D) All of the above.

104. The deficiency of folate is characterized by an elevated excretion of formiminoglutamate (FIGLU), formed in the metabolism of

(A) Glutamate (B) Tryptophan
(C) Histidine (D) Arginine.

105. Histidine, proline and arginine, during the course of their metabolism produce the following amino acid

(A) Aspartate (B) Glutamate
(C) Glycine (D) Ornithine.

106. Nitric oxide is formed from the amino acid
(A) Arginine (B) Citrulline
(C) Ornithine (D) Histidine.

107. Nitric oxide performs the following function(s)
(A) Acts as a vasodilator
(B) Inhihits platelet aggregation
(C) Involved in the erection of penis
(D) All of the above.

108. The precursor for the synthesis of carnitine
(A) Lysine (B) Methyllysine
(C) Dimethyllysine (D) Trimethyllysine.

109. The following statement(s) on lysine metabolism are true
(A) It does not participate in transamination
(B) It is involved in carnitine synthesis
(C) It is a ketogenic amino acid
(D) All of the above.

110. Glutamate is important in the clotting factors (II, VII, IX, X) as
(A) Hydroxyglutamate (B) Carboxylglutamate
(C) Methyglutamate (D) Glutamate.

111. The following statements on γ-carboxy-glutamate are true except
(A) It is synthesized from glutamate
(B) Formation requires coenzyme pyridoxal phosphate
(C) It is an excitatory neurotransmitter
(D) Forms a bypass route for the entry of glutamate to TCA cycle.

112. The amino acid glutamine is involved in
(A) Donating nitrogen atom for purine and pyrimidine synthesis
(B) Transport and storage of ammonia
(C) Conjugation reactions of detoxification
(D) All of the above.

113. Aspartate performs the following functions except it
(A) Donates one amino group for the synthesis of urea
(B) Forms a connecting link between urea cycle and TCA cycle
(C) Is utilized for the synthesis of purines and pyrimidines
(D) Participates in the acid-base balance of the body.

114. The following amino acid shoulders the responsibility of ammonia transport in blood after glutamate and glutamine
(A) Glycine (B) Alanine
(C) Phenylalanine (D) Aspartate.

115. Serine is involved in
(A) One carbon metabolism
(B) Synthesis of glycine and alanine
(C) Synthesis of ethanolamine and choline
(D) All of the above.

116. The carbon skeleton of the amino acids has the following fate(s)
(A) Generation of energy
(B) Synthesis of glucose or fat
(C) Formation of non-essential amino acids
(D) All of the above.

117. The following amino acids are both glycogenic and ketogenic except
(A) Isoleucine (B) Valine
(C) Tyrosine (D) Tryptophan.

118. An example for only ketogenic amino acid
(A) Leucine (B) Valine
(C) Isoleucine (D) Tyrosine.

119. Glycogenic amino acids can form the following metabolites except
(A) Pyruvate (B) Succinyl CoA
(C) Acetyl CoA (D) Oxaloacetate.

120. The amino acids forming the following metabolites are ketogenic
(A) α-ketoglutarate and succinyl CoA
(B) Fumarate and oxaloacetate
(C) Pyruvate and α-ketoglutarate
(D) Acetoacetate and acetyl CoA.

121. The carbon skeleton of the following amino acids can be converted to pyruvate except
(A) Glycine and alanine
(B) Aspartate and glutamate
(C) Serine and threonine
(D) Cysteine and hydroxyproline.

122. The precursors for the synthesis of polyamines
(A) Ornithine and methionine
(B) Citrulline and S-adenosylmethionine
(C) Ornithine and S-adenosylmethionine
(D) Arginine and S-adenosylmethionine.

123. The following statements on the biosynthesis of polyamines are true except
(A) S-adenosylmethionine donates methyl group
(B) Ornithine decarboxylase is the rate limiting enzyme
(C) The non-protein amino acid ornithine is required
(D) S-adenosylmethionine donates carbon moiety.

124. One of the following amino acids has been implicated in the risk for coronary heart disease

(A) Methionine (B) Cysteine

(C) Homocysteine (D) None of the above.

125. Diagnostically, the ideal marker(s) among the polyamines for cell proliferation

(A) Spermine (B) Spermidine

(C) Putrescine (D) All of the above.

126. A 65 year old man had muscular rigidity, tremors, expressionless face, lethargy, involuntary movement etc. His physician prescribed the drug-L-DOPA along with α-methyl-DOPA. What is the diagnosis?

(A) Gout (B) Parkinson's disease

(C) Scurvy (D) Pellagra

127. A patient a malignant carcinoid syndrome will develop the symptoms of

(A) Anemia (B) Pellagra

(C) Nyctalopia (D) Rickets

128. Hyperhomocysteinemia is associated with increased risk of thrombosis and coronary heart diseases. The following vitamin supplemenation reduces hyperhomocysteinemia

(A) Thiamine (B) Niacin

(C) Pyridoxine (D) Folic acid

Answers

1. **(B)** 2. **(C)** 3. **(D)** 4. **(A)** 5. **(D)** 6. **(B)** 7. **(C)** 8. **(C)** 9. **(B)** 10. **(A)** 11. **(C)** 12. **(D)** 13. **(C)** 14. **(A)** 15. **(D)** 16. **(D)** 17. **(D)** 18. **(C)** 19. **(D)** 20. **(C)** 21. **(D)** 22. **(C)** 23. **(A)** 24. **(B)** 25. **(C)** 26. **(D)** 27. **(D)** 28. **(B)** 29. **(D)** 30. **(D)** 31.**(D)** 32. **(A)** 33. **(A)** 34. **(C)** 35. **(D)** 36. **(B)** 37. **(C)** 38.**(B)** 39. **(D)** 40. **(A)** 41. **(B)** 42. **(A)** 43. **(B)** 44. **(C)** 45. **(C)** 46. **(B)** 47. **(A)** 48. **(B)** 49. **(B)** 50. **(C)** 51. **(B)** 52.**(B)** 53. **(B)** 54. **(A)** 55. **(B)** 56. **(A)** 57. **(D)** 58. **(C)** 59. **(B)** 60. **(C)** 61. **(B)** 62. **(D)** 63. **(B)** 64. **(A)** 65. **(D)** 66. **(B)** 67. **(B)** 68. **(C)** 69. **(D)** 70. **(C)** 71. **(B)** 72. **(A)** 73. **(C)** 74. **(C)** 75. **(B)** 76. **(C)** 77. **(A)** 78. **(A)** 79. **(D)** 80. **(C)** 81. **(D)** 82. **(D)** 83. **(C)** 84. **(B)** 85. **(D)** 86. **(A)** 87. **(B)** 88.**(C)** 89. **(C)** 90. **(D)** 91. **(B)** 92. **(D)** 93. **(D)** 94. **(A)** 95. **(B)** 96. **(D)** 97. **(B)** 98. **(C)** 99. **(D)** 100. **(C)** 101. **(B)** 102. **(C)** 103. **(D)** 104. **(D)** 105. **(B)** 106. **(A)** 107. **(D)** 108. **(D)** 109. **(D)** 110. **(B)** 111. **(C)** 112. **(D)** 113. **(D)** 114. **(B)** 115. **(D)** 116. **(D)** 117. **(B)** 118. **(A)** 119. **(C)** 120. **(D)** 121. **(B)** 122. **(C)** 123. **(A)** 124. **(C)** 125. **(C)** 126. **(B)** 127. **(B)** 128. **(D)**.

Clinical Case Study 15.1

(Source : Biochemistry, Page 337)

A six month old male infant had growth retardation, lethargy, frequent vomitings. His plasma ammonia concentration was 520 µg/dl (normal 15-45 µg/dl). He also had elevated plasma glutamate concentration and reduced citrulline. Urine analysis revealed increased excretion of glutamine and uracil.

1. **What is the probable diagnosis?**
2. **Name the enzyme defect.**
3. **What is the biochemical explanation for increase in plasma ammonia?**
4. **Describe the transport and storage of ammonia.**

Ans. :

1. **Hereditary hyperammonemia**.
2. Ornithine transcarbamoylase.
3. Due to a blockage in the urea cycle, carbamoyl phosphate which accumulates in the mitochondria leaks out into cytosol to be utilized for increased uracil formation (hence elevated uracil excretion in urine). Glutamate and glutamine are the carriers of ammonia, therefore they get elevated in circulation as well as urine.
4. Refer Chapter 15, page 337.

Clinical Case Study 15.2

(Source : Biochemistry, Page 347)

A 35-year old man was operated thrice for renal stones in a span of 5 years. The chemical analysis of the stones revealed that they were primarily composed of oxalate. The man's urine contained elevated oxalate concentration — 140 mg/day against normal <60 mg/day.

1. **What is the probable diagnosis?**
2. **What is the enzyme defect?**
3. **What is the biochemical basis of increased oxalate?**
4. **Name the biologically important products of glycine metabolism.**

Ans. :

1. **Primary hyperoxaluria**
2. Glycine transaminase
3. Due to a metabolic blockade in glycine metabolism, glyoxylate accumulates and gets converted to oxalate. Elevated urine excretion of oxalate causes frequent stone formation.
4. Creatine, heme, glutathione, purines, conjugation.

Clinical Case Study 15.3
(Source : Biochemistry, Pages 352-353)

A 5-year old boy had delayed developments. He was found to be mentally retarded with characteristic seizures and tremors. The boy had a plasma phenylalanine level of 30 mg/dl (reference range of 1-2 mg/dl). His urine gave a positive ferric chloride test, indicating an elevated excretion of phenylpyruvate.

1. **What is the probable diagnosis?**
2. **What is the enzyme defect?**
3. **What is the metabolic fate of phenylalanine in the disorder?**
4. **What is the mode of treatment?**

Ans. :

1. **Phenylketonuria (PKU)**
2. Phenylalanine hydroxylase.
3. Phenylalanine cannot be converted to tyrosine, hence gets diverted to alternate pathways, producing metabolites which are directly or indirectly inolved in growth failure and mental retardation.
4. The treatment for PKU is to provide diet low or deficient in phenylalanine, so that their plasma phenylalanine levels are kept within the normal limits.

Clinical Case Study 15.4
(Source : Biochemistry, Pages 353-354)

A 50-year old man had a complaint of arthritis. His urine gave a positive Benedict's test for reducing substances. On careful examination, his urine was found to turn to blackish (coke-like) colour on standing. The laboratory investigations revealed that the old man had normal blood glucose and serum uric acid (4 mg/dl) concentrations.

1. **What is the probable diagnosis?**
2. **Name the metabolism and enzyme defect.**
3. **What is the metabolite excreted into urine responsible for blackish colour?**
4. **What is the explanation for arthritis in this disorder?**

Ans. :

1. **Alkaptonuria**
2. Inborn error of tyrosine metabolism, due to a defect in the enzyme homogentisate oxidase.
3. Homogentisate — a strong reducing substance (hence Benedict's test positive). Homogentisate on standing gets oxidized to impart the characteristic coke colour to urine.
4. Arthritis in alkaptonuric patients is believed to be due to the deposition of pigment alkapton (in the joints) produced from homogentisate.

Clinical Case Study 15.5
(Source : Biochemistry, Pages 354-355)

A full term baby, born to normal and healthy parents, was found to have marked lack of pigmentation. The baby had white skin, hair and blue eyes. The paternal grandfather of the baby had similar problem.

1. **What is the pigment lacking, and what is the probable diagnosis?**
2. **Name the enzyme defect.**
3. **What are the characteristics of this disorder?**

Ans. :

1. The generalized lack of pigmentation (**melanin**) of the baby is good example of **albinism**.
2. The enzyme tyrosinase that is mainly responsible for the formation of melanin.
3. The albinos are sensitive to light and are susceptible to skin cancers. Photophobia (intolerance to light) is associated with lack of pigmentation in eyes.

Clinical Case Study 15.6
(Source : Biochemistry, Pages 363-364)

A 12-year old boy was brought to the hospital with a complaint of severe pain in his lower right flank. His urine was found to have elevated levels of cystine, lysine, arginine and ornithine. Microscopic examination of urine revealed the presence of hexagonal crystals. Further, X-rays showed the occurrence of stones in the kidney.

1. **What is the probable diagnosis?**
2. **What is the defect?**
3. **What is the explanation for stones in kidney and urinary tract?**
4. **What is the mode of treatment?**

Ans. :

1. **Cystinuria**, an inborn error of sulfur containing amino acids.
2. A defect in the carrier system in the renal tubules that is responsible for the absorption of cystine, ornithine, arginine and lysine, their excretion is increased in urine.
3. Cystine is relatively insoluble and therefore gets precipitated leading to the formation of cystine stones in the kidney and urinary tract.
4. Restricted ingestion of dietary cystine and high intake of fluids.

Clinical Case Study 15.7
(Source : Biochemistry, Page 364)

A 5-year old girl was found to have stunted growth, delayed developments and mental retardation. X-ray indicated that she had signs of osteoporosis. Her urine gave a positive test for homocysteine. Vitamin B_6 supplementation did not make any improvement.

1. **What is the probable diagnosis?**
2. **What is the enzyme defect, and blocked metabolic reaction?**
3. **Describe different types of this disorder with defective enzymes.**

Ans. :

1. **Homocystinuria.**
2. The biochemical lesion lies in the deficiency of the enzyme cystathionine synthase, as a result of which homocysteine cannot be converted to cystathionine.
3. Refer Chapter 15, pages 364-365.

Clinical Case Study 15.8
(Source : Biochemistry, Page 367)

A 2-week old infant was passing urine which smelled like burnt sugar. The plasma and urine concentrations of branched chain amino acids (valine, leucine and isoleucine) were found to be very high.

1. **Name the disorder along with the metabolism it is linked to.**
2. **What is the enzyme defect?**
3. **What is the mode of treatment?**
4. **What happens if not treated in time?**

Ans. :

1. **Maple syrup urine disease (MSUD)**, an inborn error of branched chain amino acid metabolism.
2. Branched chain α-keto acid dehydrogenase which catalyses the oxidative decarboxylation of branched chain amino acids.
3. Diet low in branched chain amino acids.
4. If not corrected, the disease results in acidosis, lethargy, convulsions, mental retardation, coma and finally death within one year after birth.

For more clinical case studies, refer 7.5, 7.6

VIVA/Short Questions

1. What do you understand by amino acid pool?

The total quantity of the free amino acids (approximately 100 g) present in the body. Glutamate and glutamine together constitute about 50% of the amino acid pool.

2. What are the major sources of the amino acid pool?

Intake of dietary proteins, turnover of body proteins and synthesis of non-essential amino acids.

3. How are the amino acids from the body pool utilized?

Synthesis of body proteins, biologically important nitrogenous compounds (porphyrins, purines, pyrimidines), carbohydrates and fats, besides contribution to body energy requirements (10-15%).

4. What are the common reactions of amino acid metabolism?

Transamination, followed by deamination liberating ammonia. The amino group of the amino acids is utilized for the formation of urea which is an excretory end product of protein metabolism.

5. What is transamination?

The transfer of an amino (—NH_2) group from an amino acid to a keto acid.

6. What are the characteristics of transamination?

Transamination is reversible, requiring the coenzyme, PLP. It is important for the synthesis of non-essential amino acids and generating energy.

7. What is deamination?

The removal of amino group from the amino acids as ammonia. It may be oxidative or non-oxidative in nature.

8. What do you understand by transdeamination?

The simultaneous occurrence of transamination (transfer of amino group) and deamination (removal of amino group).

9. What is oxidative deamination?

The liberation of free ammonia from the amino group of the amino acids coupled with oxidation.

10. What do you know about glutamate dehydrogenase (GDH)?

GDH catalyses the oxidative deamination of glutamate to liberate ammonia. This enzyme contains zinc and is located in mitochondria. GDH links glutamate metabolism with Krebs cycle.

11. What is non-oxidative deamination?

Some of the amino acids can be deaminated to liberate NH_3 without undergoing oxidation. e.g., serine and threonine.

12. What are different sources for ammonia formation in the body?

Amino acids, biogenic amines, amino group of purines and pyrimidines and by the action of intestinal bacteria (enzyme urease) on urea.

13. How is ammonia transported in the body?

Predominantly in the form of glutamine or alanine and not as free ammonia. Further, glutamine serves as a storage form of ammonia.

14. 'Ammonia is not just a waste product in the body'—Justify.

Ammonia is involved (directly or via glutamine) for the synthesis of several important compounds in the body. These include non-essential amino acids, purines, pyrimidines, amino sugars etc. Further, ammonium ions are very important to maintain acid-base balance in the body.

15. What are the different ways, the animals have developed, for the disposal of ammonia from the body?

The ammoniotelic animals dispose off NH_3 into the surrounding water. Uricotelic animals (reptiles and birds) convert NH_3 into uric acid while ureotelic mammals (including man) convert NH_3 to urea.

16. What are the toxic manifestation of ammonia?

Harmful to the brain, causes slurring of speech and blurring of vision, and results in tremors, coma and finally death, if not corrected.

17. What is the end product of protein metabolism and where is it synthesized?

Urea, synthesized in the liver.

18. How many amino groups are present in urea and how are they derived?

Urea has two amino (—NH_2) groups, one is derived from NH_3 and the other from aspartate.

19. What is the difference between carbamoyl phosphate synthese I and carbamoyl phosphate synthase II?

Carbamoyl phosphate synthase I (CPS I), is involved in urea synthesis. It is a mitochondrial enzyme catalysing the condensation of NH_4^+ with CO_2 to form carbamoyl phosphate, that requires N-acetylglutamate. CPS II, involved in pyrimidine synthesis is a cytosomal enzyme. It accepts amino group from glutamine and does not require N-acetylglutamate for its activity.

20. How is urea synthesis regulated?

The enzyme carbamoyl phosphate synthase I is the rate limiting step in urea synthesis. It is allosterically activated by N-acetylglutamate.

21. How are urea cycle and TCA cycle integrated?

Fumarate released in urea cycle enters TCA cycle where it is converted to malate and then to oxaloacetate. The latter on transamination forms aspartate that can enter urea cycle. In addition, ATP and CO_2 produced in TCA cycle can be utilized in urea cycle.

22. What is the diagnostic importance of blood urea estimation?

As a screening test for the evaluation of kidney (renal) function. Elevation in blood urea (normal <40 mg/dl) is associated with disorders pre-renal (diabetic coma), renal (acute glomerulonephritis) and post-renal (urinary stones) conditions.

23. What do you know about non-protein nitrogen substances?

Refers to all the nitrogen-containing substances other than proteins. These include urea (most abundant), creatinine, creatine, uric acid, peptides, amino acids etc. In healthy individuals, NPN concentration in blood is 20-40 mg/dl.

24. What is BUN and how is it related to NPN?

Blood urea nitrogen (BUN) is approximately half of the blood urea concentration.

$$BUN = \frac{1}{2} NPN; \; NPN = 2\ BUN.$$

25. What are the important metabolic functions of glycine?

Formation of specialized products such as heme, purines, creatine and glutathione.

26. How does glycine contribute for purine biosynthesis?

The entire molecule of glycine is utilized for the formation of positions 4 and 5 of carbon, and position 7 of nitrogen in purine the ring.

27. What are the functions of glycine with regard to conjugation reactions?

Glycine conjugates with bile acids to produce glycocholic acid and glycochenodeoxycholic acid. Glycine is also important for detoxification of benzoic acid to form hippuric acid.

28. What do you know about creatine biosynthesis?

The three amino acids glycine, arginine and methionine are required for creatine formation.

29. What is the diagnostic importance of serum creatinine estimation?

Elevation in serum creatinine (normal <1 mg/dl) reflects impairment in kidney functions.

30. What is creatinine coefficient?

The mg of creatinine and creatine (put together) excreted per kg body weight per day. For a normal adult man, 24-26 mg. while for a woman, 20-22 mg.

31. What do you know about the disorders of glycine metabolism?

Primary hyperoxaluria, due to a defect in the enzyme glycine transaminase, causes increased excretion of urinary oxalate.

32. What are the important metabolic functions of tyrosine or phenylalanine?

Synthesis of hormones — epinephrine, norepinephrine, thyroxine and triiodothyronine and the pigment melanin.

33. How is phenylalanine converted to tyrosine?

The enzyme phenylalanine hydroxylase converts phenylalanine to tyrosine. A defect in this ezyme results in the metabolic disorder phenylketonuria.

34. What is melanin and how is it synthesized?

Melanin is the pigment of skin, hair and eyes. It is synthesized from tyrosine by the action of the enzyme tyrosinase.

35. What is the role of thyroglobulin in the synthesis of thyroid hormones?

Thyroxine and triiodothyronine are synthesized from the tyrosine residues of the protein thyroglobulin and activated iodine.

36. What do you know about catecholamines?

The amine derivatives of catechol (dihydroxylated phenyl ring) are referred to as catecholamines. They are dopamine, norepinephrine and epinephrine, synthesized from tyrosine.

37. What is Parkinson's disease?

Characterized by muscular rigidity, tremors, expressionless face, lethargy etc. It is associated with decreased synthesis of dopamine.

38. What is phenylketonuria (PKU) and what are its manifestations?

PKU is due to a defect in the hepatic enzyme phenylalanine hydroxylase, as a result phenylalanine cannot be converted to tyrosine. PKU is characterized by the accumulation of phenylalanine and its alternate metabolites (phenylpyruvate, phenyllactate, phenylacetate) in the tissue and their excretion into urine. The other manifestation of PKU affected children include failure of growth, mental retardation and hypopigmentation of the skin.

39. What do you know about alkaptonuria?

A defect in the enzyme homogentisate oxidase in tyrosine metabolism causes alkaptonuria. As a result, homogentisate accumulates in tissues and is excreted into urine (coke colour urine due to oxidation of homogentisate).

40. What is albinism?

An inborn error, due to the lack of the synthesis of the pigment, melanin. The most common cause of albinism is a defect in the enzyme tyrosinase. Albinos are photophobic (intolerance to light) and susceptible to skin cancers.

41. What are the major metabolic functions of tryptophan?

Synthesis of NAD^+ and $NADP^+$ (coenzymes of niacin), serotonin and melatonin, besides being glycogenic and ketogenic.

42. How is serotonin synthesized and what are its functions?

Serotonin (5-hydroxytryptophan, 5 HT) is synthesized from tryptophan. It is a neurotransmitter closely involved in the control of cerebral activity (excitation) and behavioural pattern.

43. What do you know about malignant carcinoid syndrome?

A tumor caused by an uncontrolled growth of argentaffin cells of gastrointestinal tract. In this disorder, a very high amount of tryptophan (upto 60% against normal 1%) is utilized by argentaffin cells for serotonin synthesis.

44. How is melatonin synthesized and what are its functions?

Melatonin is a hormone synthesized from tryptophan in the pineal gland. Melatonin is closely connected with the circadian rhythms or diurnal variations and sleep and wake process.

45. What are the major functions of methionine?

Methionine (as S-adenosylmethionine) is utilized for transmethylation reactions, besides its conversion to cysteine and cystine.

46. What do you know about transmethylation?

The transfer of methyl group (—CH_3) from active methionine (S-adenosylmethionine) to an acceptor is known as transmethylation. e.g., synthesis creatine, epinephrine, choline, melatonin, methylated amino acids in proteins, methylated bases in tRNA.

47. What is transsulfuration?

The transfer of sulfur from one compound to another e.g., synthesis of cysteine.

48. What is active sulfate?

Active sulfate (3′-phosphoadenosine 5′-phosphosulfate, PAPS) is formed when sulfate formed in the metabolism of sulfur containing amino acids condenses with ATP. PAPS is utilized for the synthesis of mucopolysaccharides, and in detoxification.

49. What do you know about cystinuria?

Cystinuria (cystine-lysinuria) is characterized by increased urinary excretion of cystine. This is due to a defect in the carrier system in the renal tubules responsible for the reabsorption of cystine in kidney.

50. What do you know about one carbon metabolism?

The exchange of one carbon units is particularly associated with amino acid metabolism. The one carbon units include methyl ($—CH_3$), hydroxymethyl ($—CH_2OH$), methenyl ($—CH=$) and formyl ($—CH=O$) groups. Tetrahydrofolate is actively involved in one carbon metabolism.

51. What is the role of methionine and vitamin B_{12} in one carbon metabolism?

Methionine, in the form of S-adenosylmethionine, participates in transmethylation (transfer of $—CH_3$ group) reactions. For the regeneration of methionine, N^5-methyltetrahydrofolate and methylcobalamin (of B_{12}) are required.

52. What is the fate of carbon skeleton of branched chain amino acids?

The carbon skeleton of valine is glycogenic, leucine is ketogenic, while that of isoleucine is both glycogenic and ketogenic.

53. What is maple syrup urine disease?

An inborn error of branched chain amino acids due to a defect in the enzyme α-keto acid dehydrogenase.

54. What do you know about FIGLU?

Formiminoglutamate (FIGLU) is produced in histidine metabolism. It donates formimino group to tetrahydrofolate to participate in one carbon metabolism. Deficiency of folic acid causes elevated urinary excretion of FIGLU.

55. What are the functions of nitric oxide (NO)?

NO acts as a vasodilator causing relaxation of smooth muscles, inhibits platelet aggregation and adhesion, functions as a neurotransmitter and is involved in the erection of penis.

56. How is carnitine synthesized?

Carnitine is synthesized from trimethyllysine.

57. What do you know about γ-aminobutyric acid?

It is formed from glutamate by the action of a pyridoxal phosphate dependent glutamate decarboxylase. It is an important inhibitory neurotransmitter in the brain. Decrease in GABA levels will cause convulsions.

58. What is the metabolic importance of glutamine?

Glutamine acts as a storage and transport form of ammonia. It is the donor of nitrogen atom for purine and pyrimidine synthesis. Glutamine takes part in conjugation reactions and acid-base balance.

59. What is the fate of carbon skeleton of amino acids?

Oxidation via TCA cycle to produce energy, synthesis of glucose, formation of lipids, synthesis of non-essential amino acids.

60. Name metabolic products formed on the degradation of carbon skeleton of amino acids?

Pyruvate, α-ketoglutarate, succinyl CoA, fumarate, oxaloacetate, acetyl CoA and acetoacetate.

61. How are the amino acids classified, based on the nature of metabolic end products of carbon skeleton?

Glycogenic (glucogenic), ketogenic or both glycogenic and ketogenic.

62. What are polyamines and what is their biochemical importance?

Polyamines possess multiple amino groups, e.g., putrescine, spermine, spermidine. They are involved in the biosynthesis of DNA, RNA and proteins. The urinary excretion of polyamines is found to be elevated in almost all types of cancers.

16. Integration of Metabolism

MCQs

1. **The metabolic pathway of integration**
   (A) Embden-Meyerhoff pathway
   (B) Electron transport chain
   (C) Krebs cycle
   (D) Fatty acid oxidation.
2. **In the following pathway, the dietary foodstuffs get finally oxidized to CO_2**
   (A) Hexose monophosphate shunt
   (B) Glycolysis (C) Citric acid cycle
   (D) Urea cycle.
3. **The compound that effectively serves as a fuel reserve to meet the body needs, for brief periods (particularly between meals)**
   (A) Glucose (B) Glycogen
   (C) Fat (D) Amino acid.
4. **The metabolic pathways, in general are controlled by**
   (A) The availability of substrates
   (B) Allosteric regulation of enzymes
   (C) Regulation of enzyme synthesis
   (D) All of the above.
5. **The organ that is regarded as the 'body's central metabolic clearing house'**
   (A) Liver (B) Brain (C) Heart (D) Kidney.

6. The energy storage organ/tissue of the body

(A) Liver (B) Adipose tissue
(C) Brain (D) Heart.

7. The organ that consumes the highest quantity of glucose at rest for its energy demands

(A) Heart (B) Skeletal muscle
(C) Brain (D) Adipose tissue.

8. One of the following processes at rest consumes the highest quantity of energy

(A) Activation of glucose
(B) Activation of fatty acids
(C) Na^+-K^+ pump
(D) Aspartate -malate shuttle.

9. The following compound is exported from the skeletal muscle

(A) Glucose (B) Ketone bodies
(C) Fatty acids (D) Lactate.

10. Starvation is associated with

(A) No alteration in insulin and glucagon levels
(B) A decrease in insulin and an increase in glucagon
(C) An increase in insulin and a decrease in glucagon
(D) A decrease in both insulin and glucagon.

11. The survival time of an individual on starvation mostly depends on

(A) Liver glycogen
(B) Skeletal muscle protein
(C) Adipose tissue fat
(D) Cardiac muscle protein.

12. The blood glucose buffering organ

(A) Liver (B) Skeletal muscle
(C) Heart (D) Brain.

13. Excessive fatty acid oxidation leads to the accumulation of ketone bodies in the body. This is due to the fact that

(A) Ketone bodies are not properly utilized
(B) There occurs a disturbance in fatty acid synthesis
(C) TCA cycle cannot cope up with the over production of acetyl CoA
(D) there is an impairment in the synthesis of glucose from ketone bodies.

14. The ketone body that serves most effectively as a fuel source for the peripheral tissues

(A) Acetone (B) Acetoacetate
(C) Acetyl CoA (D) β-hydroxybutyrate.

15. During the first two weeks of starvation, the brain is mostly dependent on the glucose, supplied by liver gluconeogenesis predominantly from the raw material(s)

(A) Glycogen (B) Fat
(C) Amino acids (D) All of the above.

Answers

1. **(C)** 2. **(C)** 3. **(B)** 4. **(D)** 5. **(A)** 6. **(B)** 7. **(C)** 8. **(C)** 9. **(D)** 10. **(B)** 11. **(C)** 12. **(A)** 13. **(C)** 14. **(D)** 15. **(C)**.

Clinical Case Study 16.1
(Source : Biochemistry, Pages 385-388)

A 28-year old working obese woman, was disappointed, as her marriage could not be arranged. Her friends noticed that she was very weak and lethargic. Her blood glucose was in the normal range. She had an elevated plasma alanine and ketone bodies-accetoacetate and β-hydroxybutyrate. Rothera's test for urine was positive.

1. **What is the probable diagnosis? Justify.**
2. **What are ketone bodies?**
3. **What happens to fatty acid metabolism in this condition?**
4. **What is the importance of ketone bodies to the brain?**

Ans. :

1. This is a clear case of **starvation**, as is evident from the elevation in alanine and ketone bodies in the circulation and their increased excretion in urine (Rothera's test positive).
2. Acetone, acetoacetate and β-hydroxybutyrate.
3. During starvation, due to lack of glucose supply for energy, fatty acids are excessively broken down. This is accompained by overproduction of ketone bodies.
4. Ketone bodies (not acetone) supply energy to the brain in this situation.

For more clinical case studies, refer Metabolism Chapters 13, 14, 15

VIVA/Short Questions

1. How are the metabolic pathways regulated?

The availability of substrates, the covalent modification of enzymes, allosteric regulation of enzymes, and regulation in enzyme synthesis.

2. What is the role of liver in metabolism?

The liver is specialized to serve as the body's central metabolic clearing house. It processes and distributes the nutrients to different tissues for utilization. The liver is very actively involved in almost all the metabolisms (carbohydrate, lipid, protein, nucleic acid).

3. 'Adipose tissue is the energy storehouse of the body'—Justify.

An adult human body contains about 15 kg triacylglycerol equivalent to 135,000 Cal, predominantly stored in the adipose tissue. Hence, the adipose tissue is appropriately regarded as the energy storehouse of the body.

4. How are the energy demands of brain met?

In the normal circumstances, glucose is the only fuel source to the brain. During prolonged starvation, brain predominantly depends on ketone bodies for energy supply.

5. What are the major metabolic changes in starvation?

The body's metabolism is reorganized to overcome starvation. Triacylglycerol (fat) of adipose tissue is predominant energy supplier. In addition, the structural proteins also come for the rescue to meet the fuel demands.

6. How does liver supply glucose in starvation?

Liver is the blood glucose buffering organ. During starvation, increased gluconeogenesis and elevated glycogen degradation furnish glucose to the needy tissues (mostly brain).

7. What is the influence of starvation on lipid metabolism?

Fatty acid oxidation is increased during starvation, with an elevated synthesis of ketone bodies, which serve as the fuel source for the peripheral tissues.

8. What is the influence of starvation on brain metabolism?

During the first 2 weeks of starvation, the brain is mostly dependent on glucose, supplied through gluconeogenesis from the amino acids. The brain slowly adapts itself to utilize ketone bodies.

17. Metabolism of Nucleotides

MCQs

1. The parent nucleotide in the biosynthesis of purine nucleotides

(A) AMP (B) GMP (C) XMP (D) IMP.

2. The purines are built upon a pre-existing

(A) Ribose 5-phosphate
(B) Deoxyribose 5-phosphate
(C) Phosphoribosyl pyrophosphate
(D) Ribulose 5-phosphate.

3. The following tissues/organs cannot synthesize purine rings of ribonucleotides

(A) Liver and kidney
(B) Brain and erythrocytes
(C) Intestine and muscle
(D) Heart and adipose tissue.

4. Ribose 5-phosphate that participates in the synthesis of purine ribonucleotides is predominantly produced in

(A) Glycolysis
(B) Hexose monophosphate shunt
(C) Uronic acid pathway
(D) Citric acid cycle.

5. Incorporation of CO_2 (carboxylation) in the synthesis of purine ring is dependent on

(A) Biotin (B) Folic acid
(C) Vitamin B_{12} (D) None of the above.

6. One of the following vitamins is closely associated with the biosynthesis of purine ribonucleotides

(A) Thiamine (B) Riboflavin
(C) Folic acid (D) Biotin.

7. The one carbon fragment that is transferred in purine nucleotide biosynthesis

(A) Formyl (B) Methyl
(C) Methenyl (D) Formimino.

8. The amino acids involved in purine ring formation

(A) Aspartate, glutamate and glycine
(B) Asparagine, glutamate and glycine
(C) Aspartate, glutamine and glycine
(D) Aspartate, glutamine and cysteine.

9. One of the following enzymes is important in the salvage pathway of purine nucleotide biosynthesis

(A) Xanthine oxidase
(B) Hypoxanthine-guanine phosphoribosyl transferase
(C) Nucleotidase
(D) PRPP glutamyl amidotransferase.

10. The intracellular concentration of the following precursor largely regulates purine nucleotide biosynthesis

(A) Ribose 5-phosphate (B) Glutamine

(C) Aspartate

(D) None of the above.

11. The enzyme involved in the feedback regulation of purine nucleotide biosynthesis

(A) PRPP synthase

(B) PRPP glutamyl amidotransferase

(C) Cyclohydrolase

(D) Adenosuccinate lyase.

12. The reducing equivalents for the conversion of purine (or pyrimidine) ribonucleotides to deoxyribonucleotides are supplied by

(A) Tetrahydrobiopterin (B) Thioredoxin

(C) $FADH_2$ (D) $NADPH + H^+$.

13. The excretory end product(s) of purine metabolism in humans

(A) Uric acid (B) Urea

(C) Allantoin (D) All of the above.

14. The enzyme xanthine oxidase contains

(A) FAD, zinc and iron

(B) FAD, molybdenum and iron

(C) NAD^+, molybdenum and iron

(D) $NADP^+$, molybdenum and iron.

15. Gout is a metabolic disease associated with the overproduction of

(A) Urea (B) Uric acid

(C) Purines (D) Ammonia.

16. Primary gout is associated with the following metabolic defect(s)

(A) Lack of feedback control of PRPP synthase and/or PRPP glutamyl amidotransferase

(B) Deficiency of hypoxanthine-guanine phosphoribosyl transferase

(C) Deficiency of glucose 6-phosphatase

(D) Any one of the above.

17. The drug used for the treatment of gout

(A) Sulfonamide (B) Allopurinol

(C) Methotrexate (D) Reserpine.

18. The deficiency of the enzyme hypoxanthine-guanine phosphoribosyl transferase causes

(A) Immunodeficiency (B) Orotic aciduria

(C) Lesch-Nyhan syndrome

(D) Jaundice.

19. The following statement(s) on Lesch-Nyhan syndrome is/are true

(A) It occurs in males only

(B) It is due to a disturbance in the salvage pathway of purine synthesis

(C) It is due to a defect in the enzyme HGPRT

(D) All of the above.

20. The disease severe combined immunodeficiency (SCID) is due to a defect in the enzyme

(A) HGPRT (B) Adenosine deaminase

(C) Xanthine oxidase

(D) Purine nucleotide phosphorylase.

21. The following precursors are needed for the synthesis of pyrimidine ring

(A) Aspartate, glutamine and CO_2

(B) Asparagine, glutamate and CO_2

(C) Aspartate, glutamate and CO_2

(D) Glycine, glutamine and CO_2.

22. The pyrimidine ring is first synthesized and then attached to

(A) Phosphoribosyl pyrophosphate

(B) Ribose 5-phosphate

(C) Ribulose 5-phosphate

(D) Xylulose 5-phosphate.

23. More than half of the pyrimidine ring is contributed by

(A) Glutamine (B) Glycine

(C) CO_2 (D) Aspartate.

24. The following statements about carbamoyl phosphate synthase II are true except it

(A) Is a mitochondrial enzyme

(B) Transfers amide nitrogen to CO_2

(C) Is activated by ATP and inhibited by UTP

(D) Is a regulatory enzyme for pyrimidine synthesis in animals.

25. In prokaryotes, the enzyme carbamoyl phosphate synthase is involved in the biosynthesis of pyrimidines, besides the amino acid

(A) Histidine (B) Arginine

(C) Tryptophan (D) Phenylalanine.

26. In microorganisms, feedback regulation of pyrimidine nucleotide biosynthesis occurs through the enzyme

(A) Carbamoyl phosphate synthase II

(B) Aspartate transcarbamoylase

(C) Dihydrooratase

(D) Orotate phosphoribosyl transferase.

27. In animals, the synthesis of pyrimidine nucleotides is regulated by the enzyme

(A) Carbamoyl phosphate synthase II

(B) Aspartate transcarbamoylase

(C) Orotate phosphoribosyl transferase

(D) OMP decarbosylase.

28. Pyrimidine bases are degraded to the following two compounds

(A) Alanine and β-aminoisobutyrate

(B) β-alanine and β-aminoisobutyrate

(C) β-alanine and α-aminoisobutyrate

(D) Urea and uric acid.

29. Orotic aciduria is due to a defect in the enzyme(s)

(A) Dihydroorotate dehydrogenase

(B) Carbamoyl phosphate synthase II and dihydrooratase

(C) Orotate phosphoribosyltransferase and OMP decarboxylase

(D) Aspartate transcarbamoylase.

30. The following protein serves as a coenzyme for the conversion of ribonucleotides to deoxyribonucleotides.

(A) Thioredoxin (B) Thermogenin

(C) Myoglobin (D) Ceruloplasmin.

31. A patient found to be a victim of gout had an elevated serum uric acid level. The physician prescribed allopurinol for treatment. Which metabolism is affected in this patient?

(A) Synthesis of purine nucleotides

(B) Degration of purine nucleotides

(C) Salvage pathway of purines

(D) Uronic acid pathway.

Answers

1. **(D)** 2. **(A)** 3. **(B)** 4. **(B)** 5. **(D)** 6. **(C)** 7. **(B)** 8. **(C)** 9. **(B)** 10. **(A)** 11. **(B)** 12. **(B)** 13. **(A)** 14. **(B)** 15. **(B)** 16. **(D)** 17. **(B)** 18. **(C)** 19. **(D)** 20. **(B)** 21. **(A)** 22. **(B)** 23. **(D)** 24. **(A)** 25. **(B)** 26. **(B)** 27. **(A)** 28. **(B)** 29. **(C)** 30. **(A)** 31. **(B)**.

Clinical Case Study 17.1
(Source : Biochemistry, Pages 393-399)

A 55-year old man complained severe pain in the joints. He was a non-vegetarian and consumed alcohol occasionally. His laboratory findings

Parameter	*Subject*	*Reference range*
Serum uric acid	12 mg/dl	3-7 mg/dl
Blood urea	25 mg/dl	15-40 mg/dl
Urinary uric acid	2.5 g/day	0.5-0.7 g/day
Urinary pH	4.5	5.5-7.0

The subject responded to the treatment by the drug-allopurinol.

1. **What is the probable diagnosis?**
2. **What is the metabolic defect?**
3. **What is the explanation for arthritis?**
4. **How does allopurinol act?**

Ans. :

1. Gout
2. Overproduction of uric acid, the excretory end product of purine metabolism.
3. Crystals of sodium urate get deposited in the joints causing gouty arthritis.
4. Allopurinol competitively inhibits the enzyme xanthine oxidase and lowers the formation of uric acid, hence the relief by administering this drug.

Clinical Case Study 17.2
(Source : Biochemistry, Page 399)

A 4-year old boy showed signs of learning disability and aggressive behaviour, besides pain in the joints. It was observed that he had an irresistible urge to bite his fingers and lips. The laboratory investigations revealed that the boy had serum uric acid concentration of 10 mg/dl (reference range 4-6 mg/dl).

1. **What is the probable diagnosis?**
2. **Which metabolism is linked with this disorder?**
3. **What is the enzyme defect?**
4. **What are the complecation of this disease?**

Ans. :

1. **Lesch-Nyhan syndrome.**
2. Inborn error of purine metabolism (salvage pathway). It is a sex-linked disorder affecting only males.
3. Hypoxanthine-guanine phosphoribosyl transferase (HGPRT).
4. Increased production of uric acid, often causing gouty arthritis, besides the neurological abnormalities.

Clinical Case Study 17.3
(Source : Biochemistry, Page 402)

A 2-year old male child with retarded growth was found to have anemia. This was non-responsive to treatment by iron, folic acid, pyridoxine and cobalamin. The following are his laboratory investigative data

Parameter	*Subject*	*Reference range*
Hemoglobin	7 g/dl	14-16 g/dl
Hematocrit	30 %	40-50 %
Orotic acid (urine)	800 mg/day	<2 mg/day

1. **What is the probable diagnosis?**
2. **What is the enzyme defect?**
3. **What is the suggested mode of treatment?**

Ans. :

1. Orotic aciduria.
2. Orotate phosphoribosyl transferase.
3. Feeding a diet rich in uridine and/or cytidine is an effective treatment for orotic aciduria.

VIVA/Short Questions

1. **What is the role of ribose 5-phosphate in purine nucleotide synthesis?**

The purines are built upon a pre-existing ribose 5-phosphate.

2. **What is the committed step in purine nucleotide biosynthesis?**

The reaction catalysed by the enzyme PRPP glutamyl aminotransferase.

3. **What is the first product obtained in purine nucleotide biosynthesis?**

Inosine monophosphate (IMP or inosinic acid) is the parent purine nucleotide synthesized.

4. **What is the importance of the inhibitors of purine synthesis?**

Folic acid is essential for the synthesis of purine nucleotides. The structural analogs of folic acid (e.g., methotrexate) are widely used to control cancer. They inhibit the synthesis of purine nucleotides and thus nucleic acids (thereby cell proliferation).

5. **What do you know about the salvage pathway for purines?**

The purines can be directly converted to the corresponding nucleotides. This pathway is particularly important in certain tissues like erythrocytes and brain.

6. **How is purine nucleotide biosynthesis regulated?**

The intracellular concentration of PRPP regulates purine synthesis to a large extent. In addition, the enzyme PRPP glutamyl aminotransferase is controlled by a feedback mechanism.

7. **What is the end product of purine metabolism?**

Uric acid (2,6,8-trioxypurine).

8. **What do you know about xanthine oxidase?**

An important enzyme in purine nucleotide degradation. It converts hypoxanthine to xanthine, and xanthine to uric acid.

9. **What is your understanding of gout?**

Gout is a metabolic disease due to overproduction of uric acid, causing its elevation in the serum (normal <7 mg/dl) i.e., hyperuricemia. Primary gout is an inborn error while secondary gout is related to certain disease (leukemias, psoriasis).

10. **What is the treatment for gout?**

Allopurinol is the drug of choice for the treatment of gout. It competitively inhibits the enzyme xanthine oxidase and reduces the production of uric acid.

11. **What is the enzyme defect in Lesch-Nyhan syndrome?**

Hypoxanthine - guanine phosphoribosyl transferase (HGPRT), an enzyme of purine salvage pathway.

12. **Name the enzyme defects in purine metabolism causing immunodeficiency diseases.**

Adenosine deaminase and purine nucleoside phosphorylase.

13. **How does the pyrimidine nucleotide biosynthesis differ from that of purine nucleotides?**

Pyrimidine ring is first synthesized and then attached to ribose 5-phosphate. This is in contrast to purine nucleotide synthesis wherein purine ring is built upon a pre-existing ribose 5-phosphate.

14. **Name the precursors for the synthesis of pyrimidine ring?**

Aspartate, glutamine and CO_2.

15. **What do you know about orotic aciduria?**

The deficiency of the enzyme(s) orotate phosphoribosyl transferase and OMP decarboxylase causes orotic aciduria. It is characterized by the excretion of orotic acid in urine, severe anemia and growth retardation.

18. Mineral Metabolism

MCQs

1. **The mineral found in the most abundant quantity in human body**

(A) Calcium (B) Phosphate
(C) Magnesium (D) Iron.

2. The following element is found in the least quantity in man
(A) Iodine (B) Copper
(C) Zinc (D) Cobalt.

3. One of the following is not a principal element
(A) Sodium (B) Potassium
(C) Iron (D) Sulfur.

4. The total content of calcium in an adult man is in the range of
(A) 0.5–1.0 kg (B) 1–1.5 kg
(C) 1.5–2.0 kg (D) 2.0–2.5 kg.

5. The quantity of calcium present in the skeletal tissue (bones and teeth) is approximately
(A) 70% (B) 80% (C) 90% (D) 99%.

6. Calcium, present in the following organs, is in a dynamic state and serves as a reservoir for the continuous supply of calcium to the body
(A) Bone (B) Skeletal muscle
(C) Liver (D) Kidney.

7. Calcium performs the following functions except its participation in/as
(A) Muscle contraction
(B) Bone formation (C) Enzyme activation
(D) Coenzyme.

8. Calcium acts as a second messenger for one of the following hormones in the regulation of liver glycogenolysis
(A) Epinephrine (B) Thyroxine
(C) Growth hormone (D) Cortisol.

9. The daily requirement of calcium for an adult man is about
(A) 400 mg (B) 800 mg
(C) 1,200 mg (D) 1,600 mg.

10. Calcium absorption in the duodenum is promoted by the following factors except
(A) Lactose (B) High pH
(C) Vitamin D (D) Parathyroid hormone.

11. Intestinal absorption of calcium is inhibited by
(A) Oxalate (B) Free fatty acids
(C) Dietary fiber (D) All of the above.

12. Calcium absorption (by the intestine) is favoured by
(A) Low pH (B) Neutral pH
(C) High pH (D) None of the above.

13. The blood calcium is present in
(A) Plasma/serum
(B) Plasma and erythrocytes
(C) Plasma and leucocytes
(D) Erythrocytes.

14. The normal concentration of serum calcium
(A) 4.5-5.5 mg/dl (B) 4.5-5.5 mEq/l
(C) 9-11 mEq/l (D) 3-4.5 mg/dl.

15. Calcium is found in the blood as
(A) Ionized form (B) citrate bound
(C) Albumin bound (D) In all three forms.

16. Calcium in the following form is non-diffusable
(A) Ionized (B) Citrate bound
(C) Albumin bound (D) None of the above.

17. The regulation of plasma calcium (homeostasis of Ca) occurs by the action of
(A) Calcitriol (B) Parathyroid hormone
(C) Calcitonin (D) All of the above.

18. The synthesis of calcium binding protein in the intestinal cells is promoted by
(A) Calcitriol (B) Vitamin D
(C) Parathyroid hormone
(D) Calcitonin.

19. Calcitriol performs the following functions except
(A) Promotion of calcium uptake by the intestinal cells
(B) Stimulation of calcium uptake by osteoblasts
(C) Enhancing decalcification of bone
(D) None of the above.

20. The secretion of parathyroid hormone is promoted by low plasma concentration of
(A) Ca^{2+} (B) Phosphate
(C) Calcitonin (D) Calcitriol.

21. The release of parathyroid hormone from parathyroid glands is under the negative feedback control of plasma
(A) Ca^{2+} (B) Mg^{2+}
(C) Phosphate (D) Calcitonin.

22. The prime function of parathyroid hormone is to elevate plasma
(A) Calcium (B) Phosphate
(C) Vitamin D (D) Calcitonin.

23. Parathyroid hormone directly causes
(A) Decalcification of bone
(B) Increased intestinal absorption of calcium
(C) Enhanced calcium excretion by kidneys
(D) Calcification of bone.

24. In a prolonged dietary deficiency of calcium, parathyroid hormone significantly promotes the following to maintain plasma calcium concentration
(A) Absorption of calcium from the intestine
(B) Reabsorption of calcium by the renal tubules
(C) Decalcification of bone
(D) Loss of calcium from the skeletal muscle.

25. Parathyroid hormone enhances the intestinal absorption of calcium by promoting the synthesis of
(A) Vitamin D
(B) 25-hydroxycholecalciferol
(C) 24,25-dihydroxycholecalciferol
(D) 1,25-dihydroxycholecalciferol.

26. The hormone calcitonin is secreted by
(A) Parathyroid gland (B) Thyroid gland
(C) Adrenal medulla (D) Pituitary gland.

27. Calcitonin performs the following function(s)
(A) Promotes calcification of bone
(B) Increases the excretion of calcium into urine
(C) Decreases plasma calcium level
(D) All of the above.

28. The renal threshold for serum calcium
(A) 6 mg/dl (B) 8 mg/dl
(C) 10 mg/dl (D) 12 mg/dl.

29. Excretion of calcium into feces is increased in
(A) Hypoparathyroidism
(B) Hyperparathyroidism
(C) Vitamin D deficiency
(D) All of the above.

30. Hypercalcemia is associated with
(A) Hyperparathyroidism
(B) Hyperthyroidism (C) Hypothyroidism
(D) Vitamin D deficiency.

31. The disorder tetany is due to
(A) Hypercalcemia (B) Hypocalcemia
(C) Hypothyroidism
(D) Hypophosphatemia.

32. Hypoparathyroidism is associated with
(A) A decrease in serum calcium
(B) An increase in serum phosphate
(C) Reduced urinary excretion of calcium and phosphate
(D) All of the above.

33. Rickets is characterized by
(A) Low levels of vitamin D in the body
(B) Dietary deficiency of calcium and phosphate or both
(C) Increased activity of alkaline phosphatase
(D) All of the above.

34. Renal rickets is characterized by a defect in the renal tubular reabsorption of
(A) Calcium (B) Phosphate
(C) Magnesium (D) Sodium.

35. Osteoporosis is characterized by
(A) Demineralization of bone
(B) Frequent bone fractures
(C) Predominant occurrence in post menopausal women
(D) All of the above.

36. Most of the body's phosphate is found in combination with
(A) Calcium (B) Magnesium
(C) Iron (D) Proteins.

37. Phosphate is essential for
(A) Formation and utilization of high energy compounds
(B) Synthesis of nucleic acids
(C) Development of bones and teeth
(D) All of the above.

38. The following statements on the serum concentration of phosphate are false except
(A) Fasting and post-prandial concentrations are the same
(B) Fasting concentration is higher than post-prandial
(C) Fasting concentration is lower than post-prandial
(D) Calcium and phosphate concentrations are same.

39. Serum phosphate concentration is elevated in
(A) Hypoparathyroidism
(B) Rickets (C) Renal rickets
(D) Hyperparathyroidism.

40. The magnesium content of an adult body
(A) 20 g (B) 40 g (C) 60 g (D) 80 g.

41. Magnesium is predomonantly required as a cofactor for the reactions catalysed by
(A) Carboxylases (B) Hydroxylases
(C) Kinases (D) Isomerases.

42. One of the following increases the intestinal absorption of magnesium
(A) Calcium (B) Phosphate
(C) Alcohol (D) Parathyroid hormone.

43. The chief cation in the extracellular fluid
(A) Sodium (B) Potassium
(C) Calcium (D) Magnesium.

44. Sodium regulates body's acid-base balance in association with
(A) Chloride and bicarbonate
(B) Chloride and sulfate
(C) Bicarbonate and sulfate
(D) Bicarbonate and phosphate.

45. Sodium is involved in the intestinal absorption of
(A) Glucose (B) Galactose
(C) Amino acids (D) All of the above.

46. The major source of sodium to the body
(A) Common salt (B) Drinking water
(C) Rice (D) Wheat.

47. The normal concentration of serum sodium
(A) 3.5-5.5 mEq/l (B) 30-35 mEq/l
(C) 95-105 mEq/l (D) 135-145 mEq/l.

48. Sodium metabolism is mainly controlled by
(A) Aldosterone (B) Calcitriol
(C) Androgens (D) Oxytocin.

49. Decreased serum sodium concentration (hyponatremia) is associated with
(A) Addison's disease
(B) Cushing's syndrome
(C) Goiter (D) Grave's disease.

50. Cushing's syndrome is characterized by an elevation in serum concentration of
(A) Calcium (B) Sodium
(C) Potassium (D) Magnesium.

51. Potassium performs the following functions except
(A) Maintaining osmotic pressure
(B) Participating in protein biosynthesis
(C) Regulation of acid-base balance
(D) Involvement in intestinal glucose absorption.

52. The normal concentration of serum potassium
(A) 2.0-3.5 mEq/l (B) 3.5-5.0 mEq/l
(C) 5.0-6.5 mEq/l (D) 6.5-8.0 mEq/l.

53. Hemolysis of RBC causes a drastic elevation in the serum concentration of
(A) Calcium (B) Sodium
(C) Potassium (D) Chloride.

54. Increased serum potassium concentration (hyperkalemia) is observed in
(A) Cushing's syndrome
(B) Addison's disease
(C) Anemia (D) Goiter.

55. The enzyme salivary amylase is activated by
(A) Sodium (B) Chloride
(C) Potassium (D) Sulfate.

56. The normal serum chloride concentration
(A) 20-30 mEq/l (B) 45-55 mEq/l
(C) 95-105 mEq/l (D) 140-160 mEq/l.

57. The sulfur content of proteins (by weight)
(A) 0.1% (B) 1% (C) 5% (D) 10%.

58. The following compounds contain sulfur except
(A) Thiamine, lipoic acid
(B) Biotin, coenzyme A
(C) Pantothenic acid, serine
(D) Heparin, glycosaminoglycans.

59. The following sulfur containing compound participates in transmethylation reactions
(A) Methionine
(B) S-adenosylmethionine
(C) S-adenosylhomocysteine
(D) Cysteine.

60. There is no specific recommended dietary allowance for the following mineral
(A) Sodium (B) Sulfur
(C) Potassium (D) Fluorine.

61. The total content of iron in an adult body
(A) 1-3 g (B) 3-5 g (C) 5-7 g (D) 7-9 g.

62. The most predominant iron containing compound in the body
(A) Cytochrome C (B) Hemoglobin
(C) Myoglobin (D) Catalase.

63. The following compounds contain heme iron except
(A) Myoglobin (B) Tryptophan pyrrolase
(C) Catalase (D) Ferritin.

64. Electron transport chain operates through the participation of
(A) Hemoglobin (B) Myoglobin
(C) Cytochromes (D) Ferritin.

65. Hemoglobin is required for the transport of
(A) O_2 alone (B) CO_2 alone
(C) O_2 and CO_2 (D) None of the above.

66. The dietary requirement of iron per day by an adult man
(A) 10 mg (B) 20 mg
(C) 30 mg (D) 40 mg.

67. The following are rich sources of iron except
(A) Liver (B) Pulses
(C) Cereals (D) Milk.

68. In normal healthy people, the percentage of dietary iron absorbed in the intestine
(A) 10% (B) 20% (C) 30% (D) 40%.

69. The following substance(s) facilitate(s) the conversion of dietary ferric iron (Fe^{3+}) to ferrous (Fe^{2+}) form for its absorption
(A) Ascorbic acid (B) Cysteine
(C) Gastric HCl (D) All of the above.

70. Intestinal absorption of iron is increased by the following except
(A) Ascorbic acid, cysteine
(B) Peptides, amino acids
(C) Phytate, oxalate
(D) Iron deficiency anemia.

71. Intestinal absorption of iron is decreased by the following except
(A) Tea and eggs (B) High phosphate
(C) Steatorrhea (D) Iron deficiency anemia.

72. The temporary storage form of iron in the mucosal cells
(A) Ferritin (B) Hemosiderin
(C) Transferrin (D) Peroxidase.

73. The following compound brings about the oxidation of plasma ferric (Fe^{3+}) iron to ferrous (Fe^{2+}) iron
(A) Ferritin (B) Ceruloplasmin
(C) Ascorbic acid (D) Transferrin.

74. Iron binds with transferrin and gets transported in the plasma. The number of iron atoms that can bind to transferrin
(A) 2 (B) 3 (C) 4 (D) 5.

75. The body's iron, stored in liver, spleen and bone marrow, is normally in the form of
(A) Hemosiderin (B) Catalase
(C) Ferritin (D) Transferrin.

76. Iron is not lost through
(A) Urine (B) Bile
(C) Sweat (D) Any one of the above.

77. Iron deficiency results in
(A) Hemolytic anemia
(B) Pernicious anemia
(C) Microcytic anemia
(D) Megaloblastic anemia.

78. The disorder commonly observed in people with repeated blood transfusions
(A) Hemosiderosis
(B) Microcytic hypochromic anemia
(C) Bronze diabetes (D) Wilson's disease.

79. Copper is an essential component of the following enzymes
(A) Cytochrome oxidase, catalase
(B) Tyrosinase, superoxide dismutase
(C) Monoamine oxidase, uricase
(D) All of the above.

80. Copper is necessary for proper synthesis of
(A) Hemoglobin (B) Collagen
(C) Melanin (D) All of the above.

81. The copper containing protein that functions like hemoglobin in invertebrates
(A) Ceruloplasmin (B) Hemocyanin
(C) Hepatocuprein (D) Hemocuprein.

82. The normal concentration of serum ceruloplasmin
(A) 5-25 mg/dl (B) 25-50 mg/dl
(C) 50-75 mg/dl (D) 75-100 mg/dl.

83. The disorder associated with a defect in the intestinal absorption of copper
(A) Wilson's disease (B) Goiter
(C) Menkes disease (D) Hemochromatosis.

84. Wilson's disease is characterized by
(A) Deposition of copper in liver, kidney and brain
(B) Low serum levels of copper and ceruloplasmin
(C) Increased excretion of amino acids and glucose into urine
(D) All of the above.

85. Most of the body's iodine is present in
(A) Liver (B) Kidney
(C) Thyroid gland (D) Salivary gland.

86. The function of iodine is closely associated with the action of
(A) Steroid hormones (B) Thyroid hormones
(C) Pituitary hormones (D) Vitamin D.

87. The major part of body's iodine is stored as
(A) Protein bound iodine
(B) Iodothyroglobulin
(C) Metalothionein (D) Gusten.

88. The normal concentration of serum iodine
(A) 1-5 µg/dl (B) 5-10 µg/dl
(C) 10-15 µg/dl (D) 15-20 µg/dl.

89. The disorder closely associated with iodine metabolism
(A) Wilson's disease (B) Goiter
(C) Bronze diabetes
(D) Cushing's syndrome.

90. The manganese within the cells is found in close association with
(A) Proteins (B) Carbohydrates
(C) Lipids (D) Nucleic acids.

91. Zinc is an essential component of the following enzymes except
(A) Carbonic anhydrase
(B) Alkaline phosphatase
(C) Carboxypeptidase (D) Tyrosinase.

92. The following metal is necessary for the storage and secretion of insulin from the β-cells of pancreas
(A) Iron (B) Copper (C) Zinc (D) Iodine.

93. Gusten is a protein of saliva, important for taste sensation. The metal present in gusten
(A) Manganese (B) Zinc
(C) Sodium (D) Cobalt.

94. One of the following enzymes contains molybdenum
(A) Xanthine oxidase
(B) Monoamine oxidase
(C) Ascorbic acid oxidase
(D) Catalase.

95. Cobalt performs its biochemical functions as a constituent of
(A) Vitamin B_{12} (B) Clotting factors
(C) Thyroid hormones(D) Immunogobulins.

96. Fluoride prevents the development of dental caries in the following way(s)
(A) It forms a protective layer of fluoroapatite on the enamel
(B) It inhibits bacterial enzymes
(C) It decreases the production of acids
(D) All of the above.

97. Sodium fluoride prevents glycolysis by inhibiting the enzyme
(A) Hexokinase (B) Pyruvate kinase
(C) Enolase (D) Lactate dehydrogenase.

98. The dietary requirement of fluoride (particularly through drinking water)
(A) < 2 ppm (B) 2-5 ppm
(C) 5-8 ppm (D) 8-12 ppm.

99. Genu valgum is characterized by
(A) High intake of fluoride
(B) Hypercalcification of bones of limbs, pelvis and spine
(C) Crippling the individulas due to stiff joints
(D) All of the above.

100. The element that prevents lipid peroxidation
(A) Iron (B) Copper
(C) Selenium (D) Chromium.

101. Selenocysteine is an essential component of the enzyme
(A) Catalase (B) Glutathione peroxidase
(C) Superoxide dismutase
(D) Pyruvate carboxylase.

102. Selenium acts in association with the following vitamin
(A) Vitamin A (B) Vitamin E
(C) Vitamin B_1 (D) Vitamin B_{12}.

103. Selenosis is a toxic disorder due to high intake of selenium. This disorder is characterized by garlic odor in breath due to
(A) Monomethyl selenide
(B) Dimethyl selenide
(C) Trimethyl selenide
(D) Tetramethyl selenide.

104. In association with the following hormone, chromium promotes the utilization of glucose
(A) Cortisol (B) Insulin
(C) Glucagon (D) Epinephrine.

105. Chromium is involved in lipoprotein metabolism by
(A) Decreasing serum LDL and increasing HDL
(B) Increasing serum LDL and decreasing HDL
(C) Decreasing serum LDL only
(D) Increasing serum HDL only.

Answers

1. **(A)** 2. **(D)** 3. **(C)** 4. **(B)** 5. **(D)** 6. **(A)** 7. **(D)** 8. **(A)** 9. **(B)** 10. **(B)** 11. **(D)** 12. **(A)** 13. **(A)** 14. **(B)** 15. **(D)** 16. **(C)** 17. **(D)** 18. **(A)** 19. **(C)** 20. **(A)** 21. **(A)** 22. **(A)** 23. **(A)** 24. **(C)** 25. **(D)** 26. **(B)** 27. **(D)** 28. **(C)** 29. **(D)** 30. **(A)** 31.**(B)** 32. **(D)** 33. **(D)** 34. **(B)** 35. **(D)** 36. **(A)** 37. **(D)** 38. **(B)** 39. **(A)** 40. **(A)** 41. **(C)** 42. **(D)** 43. **(A)** 44. **(A)** 45. **(D)** 46. **(A)** 47. **(D)** 48. **(A)** 49. **(A)** 50. **(B)** 51. **(D)** 52. **(B)** 53. **(C)** 54. **(B)** 55. **(B)** 56. **(C)** 57. **(B)** 58. **(C)** 59.**(C)** 60. **(B)** 61. **(B)** 62. **(B)** 63. **(D)** 64. **(C)** 65. **(C)** 66. **(A)** 67. **(D)** 68. **(A)** 69. **(D)** 70. **(C)** 71. **(D)** 72. **(A)** 73. **(B)** 74. **(A)** 75. **(C)** 76. **(A)** 77. **(C)** 78. **(A)** 79. **(D)** 80. **(D)** 81. **(B)** 82. **(B)** 83. **(C)** 84. **(D)** 85. **(C)** 86. **(B)** 87. **(B)** 88. **(B)** 89. **(B)** 90. **(D)** 91. **(D)** 92. **(C)** 93. **(B)** 94. **(A)** 95. **(A)** 96. **(D)** 97. **(C)** 98. **(A)** 99. **(D)** 100. **(C)** 101. **(C)** 102. **(B)** 103. **(B)** 104. **(B)** 105. **(A)**.

Clinical Case Study 18.1 (Source : Biochemistry, Pages 418-419)

A 25-year old strict vegetarian woman, complained of tiredness and appeared pale. In the recent few months, the woman had heavy and prolonged menstrual flow. Her consumption of leafy vegetables was low. The laboratory investigations showed that the woman's hematocrit was 28% (reference 40% to 50%), while her hemoglobin was 8 g/dl (reference 12-14 g/dl).

1. **What is the probable diagnosis?**
2. **What is the explanation for this disorder?**
3. **What are RDAs for this nutrient?**

Ans. :

1. **Iron deficiency anemia.**
2. A strict vegetarian diet coupled with low consumption of leafy vegetables led her to iron deficiency.
3. Adult man – 10 mg/day; Menstruating woman – 18 mg/day; Pregnant and lactating woman – 40 mg/day.

Clinical Case Study 18.2 (Source : Biochemistry, Page 420)

A school boy aged 12 years complained of abdominal pain. He was weak and tired, and had a history of behavioural disturbance and epileptic form of seizures. Clinical examination revealed that the boy had an enlarged liver. His cornea showed the presence of Kayser Fleischer ring. The laboratory data of the boy are given below.

Parameter	*Subject*	*Reference range*
Serum copper	40 µg/dl	100-200 µg/dl
Serum ceruloplasmin	5 mg/dl	25-50 mg/dl
Urinary copper	200 µg/dl	< 25 µg/dl

1. **What is the probable diagnosis?**
2. **What is the mineral deposited and in which tissues?**
3. **Describe the biochemical functions of this mineral in the body.**

Ans. :

1. **Wilson's disease.**
2. This disorder is associated with the deposition of copper in liver, kidney and brain.
3. Refer Chapter 18, page 419.

Clinical Case Study 18.3 (Source : Biochemistry, Page 412)

A 10-year old boy had mild manifestations of rickets which could not be corrected by supplementation of vitamin D as well as calcium in the diet. His laboratory findings are given below.

Parameter (serum)	*Subject*	*Reference range*
Calcium	9.5 mg/dl	9-11 mg/dl
Phosphate	2.0 mg/dl	3-4.5 mg/dl
Alkaline phosphatase	300 U/l	50-100 U/l
25-Hydroxycholecalciferol	30 ng/ml	30-40 ng/ml

1. **What is the probable diagnosis?**
2. **What is the defect in this disorder?**
3. **What are the biochemical functions of vitamin D?**

Ans. :

1. **Hypophosphatemic rickets (vitamin D resistant rickets).**
2. A sex-linked disorder caused by a defect in renal tubular reabsorption of phosphate.
3. Refer Chapter 7, pages 127-130.

For more clinical case studies, refer 7.2, 7.8, 19.3, 19.4

VIVA/Short Questions

1. What are the important functions performed by minerals?

Minerals perform several vital functions essential for the very existence of the organism. These include calcification of bone, blood coagulation, neuromuscular irritability, acid-base equilibrium, fluid balance and osmotic regulation.

2. Name some organic compounds containing minerals in their structure.

Hemoglobin (Fe), insulin (Zn), thyroxine (I), thiamine (S), vitamin B_{12} (Co).

3. How are the minerals classified?

Minerals are classified as ***principal elements*** (Ca, P, Mg, Na, K, Cl and S) and ***trace elements*** (Fe, Cu, I, Mn, Zn, Mo, Co, F, Se and Cr).

4. Name the most abundant mineral in human body.

Calcium, constituting about 2% of the body weight. The total content in an adult man is about 1.5 kg.

5. How is calcium distributed in the body?

About 99% of the body calcium is present in the bones and teeth, while 1% is found in rest of the body.

6. What are the major biochemical functions of calcium?

Essential for the development of bones and teeth, muscle contraction, blood coagulation, nerve transmission, activation of certain enzymes, as an intracellular messenger for hormonal action etc.

7. What are the dietary requirements and sources of calcium?

About 800 mg/day. Milk, milk products, beans and leafy vegetables are the good sources.

8. How is calcium present in the blood?

In the plasma (or serum) with a concentration of 9-11 mg/dl (4.5-5.5 mEq/l). About half of this is in the ionized form which is functionally the most active. The remaining is bound to proteins (mostly albumin) and citrate.

9. What are the factors that regulate plasma calcium level?

Calcitriol, parathyroid hormone (PTH) and calcitonin.

10. How does calcitriol regulate plasma calcium?

Calcitriol increases the intestinal absorption of calcium, stimulates calcium uptake by osteoblasts of bone and promotes calcification.

11. How does calcitonin influence plasma calcium?

Calcitonin promotes calcification of bone and increases excretion of calcium into urine. Thus, calcitonin has a decreasing influence on plasma calcium.

12. What do you know about hypercalcemia?

Elevation in serum calcium level (normal 9-11 mg/dl). Hypercalcemia is associated with hyperparathyroidism.

13. What is hypocalcemia?

Hypocalcemia is characterized by a fall in serum calcium (normal 9-11 mg/dl) which results in tetany.

14. What do you understand by osteoporosis?

Demineralization of bone resulting in the progressive loss of bone mass. It usually occurs in the elderly people (over 60 years), the postmenopausal women being more susceptible. Osteoporosis results in frequent bone fractures which are the major cause of disability among the elderly.

15. How is sodium distributed in the body?

Sodium is the chief cation in the extracellular fluid. About 50% of the body sodium is present in the bones, 40% in the extracellular fluids and the remaining 10% in the soft tissues.

16. What are the major biochemical functions of sodium?

Maintenance of electrolyte, water and acid-base balance. Necessary for cell permeability and intestinal absorption of glucose.

17. How is the serum concentration of sodium maintained?

The normal serum concentration of sodium is 135-145 mEq/l. The mineralocorticoids, secreted by adrenal cortex, influence sodium metabolism.

18. What are the important biochemical functions of potassium?

Principal intracellular cation, required for the maintenance of water and acid-base balance, osmotic pressure.

19. How is chloride important to the body?

Chloride is involved in the regulation of acid-base balance, fluid balance and osmotic pressure. It is also involved in the formation of HCl of gastric juice, and the activation of salivary amylase.

20. How is iron distributed in the body?

The total content of iron in an adult body is 3-5 g. About 70% of this occurs in the erythrocytes of blood as a constituent of hemoglobin. At least 5% of the body iron is present in myoglobin of muscle.

21. What are the biochemical functions of iron?

Iron exerts its biochemical functions through the compounds in which it is present. Thus, hemoglobin and myoglobin are required for the transport of O_2 and CO_2, and cytochromes for electron transport chain.

22. What are the dietary requirements and sources of iron?

An adult man requires 10 mg/day, the menstruating woman 18 mg/day, while the women during pregnancy/lactation need 40 mg/day. The good sources of iron include organ meats (liver, heart, kidney), leafy vegetables, pulses, cereals, dried fruits etc.

23. How is iron transported in the plasma?

Ferric iron binds to apo-transferrin to form transferrin, in which form iron is transported in the circulation.

24. What do you know about the storage of iron in the body?

Iron is mainly stored in the body in the form of ferritin in liver, spleen and bone marrow.

25. 'Iron is a one-way substance'—Justify.

Iron is very efficiently utilized and reutilized by the body. Iron is not excreted into urine. Iron entry into the body is controlled at the absorption level, depending on the body needs. Thus, iron is appropriately regarded as a one-way substance.

26. What do you know about iron deficiency anemia?

Iron deficiency causes microcytic hypochromic anemia with reduced blood hemoglobin levels (< 12 g/dl). This is the most prevalent nutritional disorder worldover, mostly found in growing children adolescent girls, pregnant and lactating women.

27. What is your understanding of hemosiderosis?

Excessive iron accumulation in the body, leading to the deposition of ferritin and hemosiderin.

28. What is hemochromatosis?

A rare disease in which iron is directly deposited in the tissues (liver, spleen, pancreas and skin).

29. What are the major functions of copper?

Copper is an essential component of several enzymes (catalase, cytochrome oxidase). It is required for the synthesis of hemoglobin, collagen and melanin.

30. Name the copper-containing protein that functions like hemoglobin in invertebrates.

Hemocyanin.

31. How is copper transported in the plasma?

Copper is transported in the plasma in a bound form to ceruloplasmin (95%) and albumin (5%).

32. What is Menkes disease?

Menkes disease is due to a defect in the intestinal absorption of copper. The symptoms of this disorder include anemia and depigmentation of hair.

33. What do you know about Wilson's disease?

A rare disorder due to abnormal copper metabolism. It is characterized by copper deposition in liver, kidney and lenticular nucleus of brain. A failure to synthesize ceruloplasmin and/or an impairment in the binding capacity of copper with protein, are the probable causes of Wilson's disease.

34. What are the functions of iodine in the body?

Iodine is required for the synthesis of thyroid hormones, thyroxine (T_4) and triiodothyronine (T_3).

35. How is the iodine in the body stored?

As iodothyroglobulin (a glycoprotein) in thyroid gland.

36. What do you know about the disorders of iodine metabolism?

Simple goiter is due to iodine deficiency (results in hypothyroidism), while toxic goiter related to overproduction of thyroid hormones.

37. What are the functions of manganese?

Manganese is a cofactor for several enzymes (pyruvate carboxylase, arginase). It is required for the formation of bone, synthesis of hemoglobin, mucopolysaccharides and cholesterol.

38. What do you know about the biochemical functions of zinc?

Zinc is an essential component of several enzymes (carbonic anhydrase, alcohol dehydrogenase). It is necessary for the storage and secretion of insulin.

39. What is molybdenosis?

A rare disorder caused by excessive consumption of molybdenum. Its manifestations include impairment in growth, diarrhea and anemia.

40. What do you know about the nutritional importance of cobalt?

Cobalt is only important as a constituent of vitamin B_{12}, approximately 4% by weight. The functions of cobalt are the same as that of vitamin B_{12}.

41. What are the biochemical functions of fluoride?

Fluoride prevents the development of dental caries. It is also required for the development of bones. Fluoride inhibits the activities of certain enzymes (enolase).

42. What are the disease states of fluroride?

Consumption of drinking water containing < 0.5 ppm of fluoride is associated with development of dental caries in children. Excessive intake of fluoride (> 5 ppm) causes dental fluorosis.

43. What are the biochemical functions of selenium?

It prevents lipid peroxidation and protects the cells from free radical damage. In association with vitamin E.

44. What is selenocysteine and what are its functions?

Selenocysteine is an essential component of the enzyme glutathione peroxidase. This enzyme protects the cells against damage caused by H_2O_2.

45. What are the functions of chromium?

In association with insulin, chromium promotes the utilization of glucose. It decreases serum low density lipoproteins and increase high density lipoproteins.

CLINICAL BIOCHEMISTRY AND NUTRITION IV

19. Hormones

MCQs

1. **Hormones are characterized by all of the following except they (are)**
   (A) Organic chemical messengers
   (B) Produced in minute quantities by specific glands
   (C) Control metabolic and biological activities
   (D) Directly act on the target cells.
2. **The compounds that stimulate reproductive behaviour of the organisms**
   (A) Neurotransmitters (B) Prostaglandins
   (C) Interleukins (D) Pheromones.
3. **One of the following is an example of steroid hormones**
   (A) Insulin (B) Glucagon
   (C) Epinephrine (D) Cortisol.
4. **Cyclic AMP acts as a second messenger for a majority of**
   (A) Steroid hormones
   (B) Polypeptide hormones
   (C) Neurotransmitters (D) Pheromones.
5. **The following hormones act through the second messenger cAMP except**
   (A) Growth hormone (B) Epinephrine
   (C) Glucagon (D) Calcitonin.
6. **One of the following compounds is a second messenger for hormonal action**
   (A) Monoacylglycerol (B) Diacylglycerol
   (C) Triacylglycerol (D) Lecithin.
7. **Phosphatidylinositol 4,5-bisphosphate on hydrolysis produces the second messengers**
   (A) Inositol 1,4,5-triphosphate and diacylglycerol
   (B) Inositol 4,5-bisphosphate and diacylglycerol
   (C) Inositol 5-monophosphate and diacylglycerol
   (D) Inositol 1,4,5-triphosphate and monoacylglycerol.
8. **The inorganic element that can function as a second messenger for hormonal action**
   (A) Sodium (B) Magnesium
   (C) Calcium (D) Iron.
9. **The following hormones are synthesized by the hypothalamus except**
   (A) Growth hormone—releasing hormone
   (B) Growth hormone—release inhibiting hormone
   (C) Thyroid stimulating hormone
   (D) Prolactin release—inhibiting hormone.
10. **The master endocrine organ**
    (A) Hypothalamus (B) Anterior pituitary
    (C) Posterior pituitary (D) Adrenal cortex.
11. **The growth related effects of growth hormone are mediated through**
    (A) Cyclic AMP (B) Diacylglycerol
    (C) Somatomedin C (D) Calcium.
12. **The disorders gigantism in children and acromegaly in adults are due to**
    (A) Overproduction of growth hormone
    (B) Underproduction of growth hormone
    (C) Prolactin deficiency
    (D) Overproduction of gonadotropins.
13. **The hormone (in urine) employed in the laboratory for the detection of pregnancy**
    (A) Human chorionic gonadotropin
    (B) Prolactin (C) Progesterone
    (D) Estrogen.
14. **The conversion of cholesterol to pregnenolone in the adrenal cortex is promoted by**
    (A) Endorphins (B) Oxytocin
    (C) Adrenocorticotropic hormone
    (D) Gonadotropin.
15. **The natural analgesics that control pain and emotions**
    (A) Endorphins (B) Neurotransmitters
    (C) Morphine (D) Pheromones.
16. **The hormones produced by posterior pituitary (neurohypophysis)**
    (A) Oxytocin and growth hormone
    (B) Oxytocin and antidiuretic hormone
    (C) Oxytocin and prolactin
    (D) Prolactin and antidiuretic hormone.

17. The hormone that causes the contraction of pregnant uterus
(A) Estradiol (B) Vasopression
(C) Oxytocin (D) Pregnenolone.

18. The hormone primarily concerned with the regulation of water balance in the body
(A) Oxytocin (B) Prolaction
(C) Antidiuretic hormone
(D) Androgens.

19. The disorder characterized by the excretion of large volumes of urine
(A) Diabetes mellitus (B) Gout
(C) Cushing's syndrome
(D) Diabetes insipidus.

20. The hormone(s) secreted by thyroid gland
(A) Tetraiodothyronine
(B) Triiodothyronine (C) Calcitonin
(D) All of the above.

21. The precursor protein for the synthesis of thyroxine and triiodothyronine
(A) Prealbumin (B) Thyroalbumin
(C) Thyroglobulin (D) Thyroprotamine.

22. Triiodothyronine is more active than thyroxine by about
(A) 10 times (B) 20 times
(C) 30 times (D) 40 times.

23. Thyroid hormones increase the activity of the enzyme
(A) α-glycerophosphate dehydrogenase
(B) Lactate dehydrogenase
(C) Hexokinase
(D) Succinate dehydrogenase.

24. The following endocrine gland is the most susceptible for hypo- or hyperfunction
(A) Pituitary (B) Adrenal medulla
(C) Pancreas (D) Thyroid.

25. Triiodothyronine regulate(s) the production of the following hormone(s) by a feedback mechanism
(A) Thyroid stimulating hormone
(B) Thyroid stimulating hormone and thyrotropin releasing hormone
(C) Thyrotropin releasing hormone
(D) Thyrotropin releasing hormone and thyroxine.

26. Hypothyroidism in adults causes
(A) Cretinism (B) Myxoedema
(C) Cushing's syndrome
(D) Addison's disease.

27. The hormone cortisol is the most important
(A) Glucocorticoid (B) Mineralocorticoid
(C) Androgen (D) Sex hormone.

28. The following hormone is the most active and potent mineralocorticoid
(A) Androgen (B) Aldosterone
(C) Cortisol (D) Estrogen.

29. The common precursors for the synthesis of all steroid hormones
(A) Pregnenolone (B) Progesterone
(C) 11-deoxycortisol (D) Corticosterone.

30. Glucocorticoids (e.g., cortisol) bring about the following biochemical functions except
(A) Promotion of gluconeogenesis
(B) Increasing lipolysis
(C) Promotion of host's immune response
(D) Enhancement of protein biosynthesis.

31. The production of aldosterone is regulated by
(A) Renin-angiotensin
(B) Adrenocorticotropic hormone
(C) Sodium and potassium
(D) All of the above.

32. Impairment in adrenocortical function results in
(A) Addison's disease (B) Cushing's syndrome
(C) Pheochromocytoma
(D) Parkinson's disease.

33. The amino acid precursor for the synthesis of catecholamines
(A) Tyrosine (B) Tryptophan
(C) Arginine (D) Cysteine.

34. The hormones synthesized by adrenal medulla
(A) Epinephrine and norepinephrine
(B) Thyroxine and triiodothyronine
(C) Cortisol and cortisone
(D) Pregnenolone and progesterone.

35. The tumors of adrenal medulla are collectively referred to as
(A) Parkinson's diseases (B) Goiter
(C) Pheochromocytomas
(D) Cushing's syndrome.

36. The enzyme closely associated with the degradation of catecholamines
(A) Tyrosinase (B) Monoamine oxidase
(C) Thyroperoxidase (D) Pregnenolone.

37. The disease pheochromocytoma is associated with increased urinary excretion of
(A) 5-hydroxyindole acetic acid
(B) Vanillylmandelic acid
(C) Dopamine (D) Catechol.

38. The most active form of male sex hormone
(A) Testosterone
(B) Monohydrotestosterone
(C) Dihydrotestosterone
(D) Dehydrotestosterone.

39. The hormone dihydrotestosterone is involved in
(A) Growth and development of male reproductive organs
(B) Spermatogenesis
(C) Male pattern of aggressive behaviour
(D) All of the above.

40. Estrogens are involved in the following metabolic function(s)
(A) Lipogenic effect
(B) Hypocholesterolemic effect
(C) Calcification of bones
(D) All of the above.

41. The hormone essentially required for the implantation of fertilized ovum and maintenance of pregnancy
(A) Estrogens (B) Cortisol
(C) Progesterone (D) Estradiol.

42. The secretion of gastric HCl is stimulated by the gastrointestinal hormone
(A) Secretin (B) Cholescystokinin
(C) Gastric inhibitory peptide
(D) Gastrin.

43. A 30-year old woman visited her gynecologist two weeks after she missed menustral cycle. Her urine was tested and she was found to be pregnant. Which of one the following hormones was detected in urine.
(A) Follicle stimulating hormone
(B) Luteinizing hormone
(C) Human chorionic gonadotropin
(D) Human growth hormone

Answers

1. **(D)** 2. **(D)** 3. **(D)** 4. **(B)** 5. **(A)** 6. **(B)** 7. **(A)** 8. **(C)** 9. **(C)** 10. **(B)** 11. **(C)** 12. **(A)** 13. **(A)** 14. **(C)** 15. **(A)** 16. **(B)** 17. **(C)** 18. **(C)** 19. **(D)** 20. **(D)** 21. **(C)** 22. **(A)** 23. **(A)** 24. **(D)** 25. **(B)** 26. **(B)** 27. **(A)** 28. **(B)** 29. **(A)** 30. **(C)** 31. **(D)** 32. **(A)** 33. **(A)** 34. **(A)** 35. **(C)** 36. **(B)** 37. **(B)** 38. **(C)** 39. **(D)** 40. **(D)** 41. **(C)** 42. **(D)** 43. **(C)**.

Clinical Case Study 19.1
(Source : Biochemistry, Page 442)

A 30-year old woman, married five years ago, had no children. She complained of tiredness, weight gain, cold intolerance, neuromuscular pains and constipation. She was found to be anemic. Her laboratory data are given hereunder.

Parameter (serum)	*Subject*	*Reference range*
Triiodothyronine (T_3)	100 ng/dl	110-180 ng/dl
Thyroxine, total (T_4)	4 μg/dl	5-12 μg/dl
Thyroid stimulating hormone (TSH)	50 μU/ml	0.5-4 μU/ml
Cholesterol	250 mg/dl	150-200 mg/dl

1. **What is the probable diagnosis? Justify.**
2. **Is there any diagnostic importance of serum cholesterol in this disorder?**

Ans. :

1. The woman with the described clinical manifestations, elevated serum TSH and marginally lower T_4 and T_3 was a case of **hypothyroidism**.
2. Serum cholesterol has no diagnositic importance, since in serum it gets elevated in several other diseases e.g. diabetes, nephrotic syndrome).

Clinical Case Study 19.2
(Source : Biochemistry, Page 442)

A 35-year old woman complained of weight loss, despite normal or even increased appetite. She was found to have the symptoms of nervousness, irritability, anxiety, diarrhea, sweating, sensitive to heat and had increased pulse rate (tachycardia). Her laboratory findings are given here.

Parameter (serum)	*Subject*	*Reference range*
Triiodothyronine (T_3)	350 ng/dl	110-180 ng/dl
Thyroxine, total (T_4)	14 μg/dl	5-12 μg/dl
Thyroid stimulating hormone (TSH)	0-0.1 μU/ml	0.5-4 μU/ml

1. **What is the probable diagnosis? Justify.**
2. **Describe briefly the synthesis of T_3 and T_4.**

Ans. :

1. The clinical manifestations of the woman, along with reduced serum TSH and increased T_3 and T_4 indicate that the woman was suffering from **hyperthyroidism**.
2. Refer Chapter 15, Page 349 and Fig. 15.21.

Clinical Case Study 19.3
(Source : Biochemistry, Page 410)

A 50-year old post-menopausal woman had the symptoms of lethargy, depression, muscle weakness, loss of memory, loss of appetite. Over the past 10 years, she had 2 kidney stones that passed spontaneously. Chemical analysis of the stones revealed that calcium phosphate was the major constituent. The following are the laboratory findings.

Parameter (serum)	*Subject*	*Reference range*
Calcium	13 mg/dl	9-11 mg/dl
Phosphate	2.5 mg/dl	3-4.5 mg/dl
Chloride	110 mEq/l	95-105 mEq/l
Alkaline phosphatase	110 U/l	50-100 U/l

1. **What is the probable diagnosis?**
2. **Discuss the factors regulating plasma calcium lavel?**

Ans. :

1. **Hyperparathyroidism**.
2. Refer Chapter 18, Pages 409-410.

Clinical Case Study 19.4 (Source : Biochemistry, Page 411)

A 35-year old woman had the complaints of neuromuscular irritability, spasms and convulsions, besides tingling in the hands, feet and around mouth. Laboratory data showed that her serum calcium level was low (7 mg/dl against normal 9-11 mg/dl) while phosphate concentration was high (6 mg/dl against normal 3-4.5 mg/dl).

1. **What is the probable diagnosis?**
2. **What is the most likely explanation for this disorder?**

Ans. :

1. The clinical symptoms and the laboratory findings — hypocalcemia and hyperphosphatemia indicate that woman was suffering from **tetany**.
2. This might be due to hypoparathyroidism/ osteomalacia/calcium malabsorption.

Clinical Case Study 19.5 (Source : Biochemistry, Page 439)

A 50-year old woman complained of increased thirst (polydipsia) and high frequency of urination (polyuria). The specific gravity of the morning urine sample was 1.001 (reference range 1.003-1.030) while the osmolality was 200 milliosmoles/kg (reference range 500-1,200 milliosmoles/kg). A CT of hypothalamus and pituitary gland revealed a mass lesion in the hypothalamus. The woman developed tuberculoma in the brain due to tuberculosis.

1. **What is the probable diagnosis? Justify.**
2. **What is the biochemical basis of this disorder?**

Ans. :

1. The characteristic features — polyuria, polydipsia with a very dilute urine output (low specific gravity and osmolality) are indicative of the disorder **diabetes insipidus**.
2. The woman, developed tuberculoma destroying the ability of the hypothalamus to synthesize vasopressin (antidiuretic hormone) exclusively responsible for the reabsorption of water by renal tubules) ultimately resulting in diabetes insipidus.

Clinical Case Study 19.6 (Source : Biochemistry, Page 447)

A 28-year old man complained constant headache and visual disturbance. He also had the symptoms of anxiety, tremors and was found to be hypertensive. His laboratory data are given below.

Parameter	*Subject*	*Reference range*
Serum epinephrine	10 ng/ml	0.1 ng/ml
Serum norepinephrine	26 µg/l	0.5 µg/l
Urine vanillylmandelic acid (VMA)	62 mg/day	<8 mg/day

1. **What is the probable diagnosis?**
2. **What are the biochemical manifestations of this disorder?**

Ans. :

1. **Pheochromocytoma**, a tumor of adrenal medulla.
2. Elevation in the serum levels of catecholamines (epinephrine and norepinephrine) along with an increased urinary excretion of their principal metabolite - VMA.

Clinical Case Study 19.7 (Source : Biochemistry, Page 437)

A 25-year old woman visited her gynecologist as soon as she missed the menstrual cycle. The presence of human chorionic gonadotropin (hCG) hormone could be detected in her urine.

1. **What is the importance of testing for hCG in urine?**
2. **What is the outcome of the test in the woman?**
3. **Which tisue synthesizes hCG and how its levels are influenced?**

Ans. :

1. hCG in urine is used for the **detection of pregnancy**.
2. This woman was pregnant, as confired by the hormonal test.
3. hCG is synthesized by the placenta. The levels of hCG in plasma and urine increase immediately after implantation of fertilized ovum.

Clinical Case Study 19.8 (Source : Biochemistry, Page 448)

A 25-year old woman, married 3 years ago had no children. She had increased hair growth on the arms and face (like males), besides facial acne. She was not having regular menstrual periods. Laboratory investigations showed that her serum testosterone level was 2.5 ng/ml (normal <0.5 ng/ml).

1. **What is the provisional diagnosis with justification?**
2. **What is the explanation for male characteristics in this women?**

Ans. :

1. The woman had excessive hair growth as found in males and serum testosterone level was elevated. This is a case of **hirsutism**.
2. Increased serum testosterone causes male characteristics in females.

VIVA/Short Questions

1. **What are hormones?**

Chemical messengers produced in small amounts by specific tissues (endocrine glands), secreted into the blood stream to control the metabolic and biological activities in the target cells.

2. **Name the second messengers of hormonal action.**

Cyclic AMP, diacylglycerol, inositol 1,4,5-triphosphate.

3. **What are the biochemical effects of growth hormone?**

Growth hormone promotes growth through the mediation of somatomedin C. It has anabolic effect (promotes synthesis) on protein metabolism.

4. **What do you know about thyroid stimulating hormone (TSH)?**

TSH is released by the anterior pituitary. It promotes the uptake of iodide by the thyroid gland and thus the production of thyroxine (T_4) and triiodothyronine (T_3).

5. **What are the biochemical effects of endorphins and enkephalins?**

The natural analgesics that control pain and emotions. They are peptide neurotransmitters that control endogenous pain perception.

6. **Name the posterior pituitary hormones. What are their functions?**

Oxytocin and antidiuretic hormone (both nonapeptides). Oxytocin causes contraction of pregnant uterus and induces labour, besides causing milk ejection from the breast. Antidiuretic hormone (ADH, vasopressin) is primarily concerned with the regulation of water balance in the body.

7. **What do you know about diabetes insipidus?**

The excretion of large volumes of dilute urine (polyuria). It may be due to insufficient levels of ADH or a defect in the receptors of target cells.

8. **What are the hormones produced by the thyroid gland?**

Thyroxine (T_4) and triiodothyronine (T_3) which regulate the metabolic rate of the body. Thyroid gland also secretes calcitonin, a hormone concerned with calcium homeostasis.

9. **How do thyroid hormones influence the metabolism?**

Thyroid hormones (T_4, T_3) stimulate the metabolic activities and increase the oxygen consumption in most of the tissues of the body.

10. **What do you know about endemic goiter?**

Dietary iodine deficiency causes endemic goiter. It is mostly found in the geographical regions away from sea coast where the water and soil are low in iodine content.

11. **Name the hormones synthesized by adrenal cortex.**

Adrenal cortex produces about 50 steroid hormones collectively referred to as adrenocorticoids. They are classified into three groups — glucocorticoids (cortisol), mineralocorticoids (aldosterone), androgens and estrogens.

12. **What are the major functions of glucocorticoids?**

Glucocorticoids (cortisol, cortisone) promote gluconeogenesis and lipolysis.

13. **What are the actions of mineralocorticoids?**

Promote Na^+ reabsorption by the kidney. This is accompanied by excretion of K^+, H^+ and NH_4^+ ions.

14. What are the abnormalities associated with adrenocortical function?

Impairmet in adrenocortical function results in Addison's disease. Hypofunction of adrenal cortex causes Cushing's syndrome.

15. What are catecholamines?

Catecholamines are the amine derivatives of catechol nucleus (dihydroxylated phenyl ring). They are epinephrine, norepinephrine and dopamine.

16. What do you know about the biochemical functions of catecholamines?

Mobilize energy resources and prepare the individual to meet emergencies (shock, cold etc.). Epinephrine and norepinephrine increase glycogenolysis and gluconeogenesis and thus increase blood glucose level.

17. What are the abnormalities associated with catecholamine production?

Pheochromocytomas are the tumours of adrenal medulla, resulting in excessive production of epinephrine and norepinephrine that causes severe hypertension. Parkinson's disease is due to an impairment in the synthesis of dopamine by the brain.

18. How do you classify sex hormones?

Sex hormomes are categorized into 3 groups — androgens (male sex hormones), estrogens (female sex hormones) and progesterone (synthesized during menstrual cycle and pregnancy).

19. What are the functions of androgens?

The androgens, primarily dihydrotestosterone and testosterone, influence the development of secondary sexual characteristics, spermatogenesis and male pattern of aggressive behaviour.

20. What are the functions of estrogens?

Estrogens (female sex hormones) are predominantly ovarian hormones and are primarily responsible for maintenance of menstrual cycle and reproductive process in women, besides the development of female sexual characteristics.

21. What are the biochemical actions of progesterone?

Required for the implantation of fertilized ovum and maintenance of pregnancy.

22. What do you know about gastrointestinal (GIT) hormones?

GIT hormones are peptide in nature and are secreted by the specialized cells of GIT. These include gastrin, secretin, cholecystokinin and gastric inhibitory peptide.

20. Organ Function Tests

MCQs

1. Liver actively participates in the following functions except

(A) Metabolism of carbohydrates, proteins and lipids
(B) Detoxification of foreign compounds
(C) Synthesis of hormones
(D) Storage of vitamins — A, D and B_{12}.

2. The excretory functions of liver can be assessed by measuring serum

(A) Bile pigments (B) Transaminases
(C) Hippuric acid (D) Albumin.

3. The normal concentration of serum bilirubin

(A) 0.2-1.5 mg/dl (B) 1.5-3.0 mg/dl
(C) 3-4.5 mg/dl (D) 4.5-6.0 mg/dl.

4. The specific laboratory test to measure increase in serum bilirubin concentration

(A) Icterus index
(B) van den Bergh reaction
(C) Hollander's rest
(D) Fractional test meal.

5. van den Bergh reaction is direct positive due to

(A) Total bilirubin
(B) Unconjugated bilirubin
(C) Conjugated bilirubin (D) Bile salts.

6. The colour obtained in a positive van den Bergh reaction

(A) Blue (B) Green (C) Red (D) Purple.

7. van den Bergh reaction is biphasic due to the presence of

(A) Conjugated bilirubin
(B) Unconjugated bilirubin
(C) Both conjugated and unconjugated bilirubin
(D) Bile pigments and bile salts.

8. The exogenous compound employed to assess the excretory function of liver

(A) Bilirubin (B) Biliverdin
(C) Bromosulphthalein (D) Pentagastrin.

9. The enzyme most commonly employed in the laboratory to asses liver function

(A) Alanine transaminase
(B) Aspartate transaminase
(C) Alkaline phosphatase
(D) Lactate dehydrogenase.

10. The serum enzyme elevated in obstructive jaundice
(A) Alanine transaminase
(B) Aspartate transaminase
(C) Alkaline phosphatase
(D) Lactate dehydrogenase.

11. The serum enzyme elevated in alcholism
(A) γ-glutamyl transpeptidase
(B) 5′-nucleotidase (C) Alkaline phosphatase
(D) Aspartate transaminase.

12. Hepatic jaundice is characterized by elevated serum
(A) Conjugated and unconjugated bilirubins
(B) Conjugated bilirubin only
(C) Unconjugated bilirubin only
(D) Alkaline phosphatase.

13. Elevated serum conjugated bilirubin and alkaline phosphatase are observed in
(A) Hemolytic jaundice
(B) Obstructive jaundice
(C) Hepatic jaundice (D) All of the above.

14. The first enzyme/compound elevated in acute viral hepatitis
(A) Conjugated bilirubin
(B) Unconjugated bilirubin
(C) Alanine transaminase
(D) Aspartate transaminase.

15. Functional impairment in liver is frequently associated with increased synthesis of
(A) Albumin (B) Globulins
(C) Prothrombin (D) All of the above.

16. The half-life of serum albumin
(A) 10-20 days (B) 20-25 days
(C) 25-30 days (D) 30-35 days.

17. Prothrombin time can be used as a liver function test. But the deficiency of one of the following vitamins interferes with this test
(A) Vitamin A (B) Vitamin K
(C) Folic acid (D) Vitamin B_{12}.

18. Hippuric acid synthesis test is employed for measuring the liver function with regard to
(A) Detoxification (B) Synthetic function
(C) Excretory function (D) All of the above.

19. Kidneys perform the following functions except
(A) Regulation of acid-base balance
(B) Excretion of waste products
(C) Hormonal function
(D) Detoxification of foreign substances.

20. The hormone(s) synthesized by kidneys
(A) Erythropoietin (B) Calcitriol
(C) Renin (D) All of the above.

21. The term glomerular filtration rate refers to
(A) Millilitres of plasma filtered per minute by the kidneys
(B) Millilitres of plasma filtered per hour by the kidneys
(C) Millilitres of plasma filtered per second by the kidneys
(D) Litres of plasma filtered per minute by the kidneys.

22. Renal threshold for glucose
(A) 100 mg/dl (B) 180 mg/dl
(C) 280 mg/dl (D) 350 mg/dl.

23. Tubular maximum (TmG) for glucose
(A) 180 mg/dl (B) 280 mg/dl
(C) 350 mg/dl (D) 450 mg/dl.

24. The normal glomerular filtration rate
(A) 120-125 ml/min (B) 220-225 ml/min
(C) 320-325 ml/min (D) 420-425 ml/min.

25. The most sensitive test to assess renal function
(A) Blood urea estimation
(B) Serum creatinine estimation
(C) Creatinine clearance test
(D) Urea clearance test.

26. Urea clearance is less than the glomerular filtration (or creatinine clearance), since the renal tubules
(A) Partially secrete urea
(B) Partially reabsorb urea
(C) Neither secrete nor reabsorb urea
(D) Excrete urea only.

27. The following endogenous clearance is close to glomerular filtration rate
(A) Urea clearance (B) Creatinine clearance
(C) Inulin clearance (D) All of the above.

28. The maximum urea clearance
(A) 75 ml/min (B) 125 ml/min
(C) 54 ml/min (D) 150 ml/min.

29. Standard urea clearance is calculated when the urine output is
(A) Less than 2 ml/min
(B) More than 2 ml/min
(C) 2-4 ml/min (D) More than 4 ml/min.

30. The normal specific gravity of urine
(A) 1.010 (B) 1.020 (C) 1.030 (D) 1.040.

31. The normal ratio of osmolality between urine and plasma
(A) 1-2 (B) 2-4 (C) 4-6 (D) 6-8.

32. The term alkaline tide refers to an elevation in the following compound in the plasma following a meal
(A) OH^- ions (B) Bicarbonate
(C) Phosphate (D) Carbonic acid.

33. The H^+ ions of gastric HCl are derived from
(A) Carbonic acid (B) Phosphoric acid
(C) Lactic acid (D) Fumaric acid.

34. Ryle's tube is used in
(A) Liver function tests
(B) Renal tubular function tests
(C) Pancreati function tests
(D) Gastric function tests.

35. Pentagastrin employed in gastric function test is a
(A) Peptide (B) Trisaccharide
(C) Phospholipid (D) Amino acid.

36. The biological amine that is a powerful stimulant of gastric secretion
(A) Tyramine (B) Serotonin
(C) Histamine (D) Ethanolamine.

37. Hollander's test to assess gastric function employs
(A) Inulin (B) Insulin
(C) Histamine (D) Pentagastrin.

38. The most commonly employed serum enzyme for assessing pancreatic function
(A) Lipase (B) Trypsin
(C) Chymotrypsin (D) Amylase.

Answers

1. **(C)** 2. **(A)** 3. **(A)** 4. **(B)** 5. **(C)** 6. **(D)** 7. **(C)** 8. **(C)** 9. **(A)** 10. **(C)** 11. **(A)** 12. **(A)** 13. **(B)** 14. **(C)** 15. **(B)** 16. **(B)** 17. **(B)** 18. **(A)** 19. **(D)** 20. **(D)** 21. **(A)** 22. **(B)** 23. **(C)** 24. **(A)** 25. **(C)** 26. **(B)** 27. **(B)** 28. **(A)** 29. **(A)** 30. **(B)** 31. **(B)** 32. **(B)** 33. **(A)** 34. **(D)** 35. **(A)** 36. **(C)** 37. **(B)** 38. **(D)**.

Clinical Case Study 20.1
(Source : Biochemistry, Page 458)

1. An 18-year old male medical student had the complaints of loss of appetite (anorexia), nausea, headache and malaise. On clinical examination, his liver was found to be slightly enlarged. He was passing pale coloured stools but dark coloured urine. The data of his laboratory investigations

Parameter (serum)	*Subject*	*Reference range*
Bilirubin (total)	8 mg/dl	0.2-1 mg/dl
(van den Bergh reaction	Biphasic	Negative)
Conjugated bilirubin	4.5 mg/dl	0.4 mg/dl
Unconjugated bilirubin	3.5 mg/dl	0.6 mg/dl
Alanine transaminase (ALT)	180 U/l	10-45 U/l
Aspartate transaminase (AST)	80 U/l	10-50 U/l
Alkaline phosphatase (ALP)	10 U/l	50-100 U/l

1. **What is the probable diagnosis?**
2. **What are the likely causes of this disorder?**
3. **Why the stools are pale while urine is dark in this patient?**

Ans. :

1. **Hepatic jaundice.**
2. Viral infection (viral hepatitis), poisons and toxins or cirrhosis of liver.
3. The pale coloured stools are due to the absence of stercobilinogen while the dark coloured urine is due to urobilinogen.

Clinical Case Study 20.2
(Source : Biochemistry, Page 458)

A 20-year old man was treated for malaria a month ago. For the past one week, he was excreting slightly dark coloured urine and dark brown coloured stools. His laboratory findings are given hereunder.

Parameter (serum)	*Subject*	*Reference range*
Bilirubin (total)	7 mg/dl	0.2-1 mg/dl
(van den Bergh reaction	Indirect positive	Negative)
Conjugated bilirubin	0.4 mg/dl	<0.4 mg/dl
Unconjugated bilirubin	6.6 mg/dl	<0.6 mg/dl
Alanine transaminase (ALT)	30 U/l	10-45 U/l
Aspartate transaminase (AST)	40 U/l	10-50 U/l
Alkaline phosphatase (ALP)	10 U/l	50-100 U/l

1. **What is the probable diagnosis?**
2. **What is the biochemical basis of this disease?**
3. **Why the feces of these patients are dark brown in colour?**

Ans. :

1. **Hemolytic jaundice**.
2. An excessive breakdown of erythrocytes releasing hemoglobin. The heme degraded to bilirubin cannot be effectively conjugated, as a result unconjugated bilirubin in the serum was elevated.
3. Feces were dark brown due to stercobilinogen.

Clinical Case Study 20.3
(Source : Biochemistry, Page 458)

A 30-year old man had fever and abdominal pain. His stools were clay in colour and contained fat (steatorrhea). The man's urine was found to be dark in colour. His laboratory findings are given

Parameter (serum)	Subject	Reference range
Bilirubin (total)	12 mg/dl	0.2-1 mg/dl
(van den Bergh reaction	Direct positive	Negative)
Conjugated bilirubin	11.5 mg/dl	<0.4 mg/dl
Unconjugated bilirubin	0.5 mg/dl	<0.6 mg/dl
Alanine transaminase (ALT)	70 U/l	10-45 U/l
Aspartate transminase (AST)	75 U/l	10-50 U/l
Alkaline phosphatase (ALP)	450 U/l	50-100 U/l

In addition, the urine gave a positive test for bile salts (Hay's test) and bile pigments (Fouchet's test).

1. **What is the probable diagnosis?**
2. **What is the cause of this disorder?**
3. **What is the explanation for the colour changes observed in urine and stools?**

Ans. :

1. **Obstructive jaundice (cholestasis).**
2. Cholestasis is due to an obstruction (caused by gall stones, tumors etc.) that prevents the passage of bile into the intestine.
3. The urine was dark in colour due to the excretion of conjugated bilirubin while the stools were clay in colour due to the deficiency of stercobilinogen, a pigment derived from bilirubin. Feces contained excess fat indicating impairment in fat digestion and absorption due to the absence of bile sats.

Clinical Case Study 20.4
(Source : Biochemistry, Page 457)

A 50-year old man, an industrialist by profession, had a routine medical check up. He was found to have slightly enlarged liver. His laboratory findings.

Parameter	Subject	Reference range
Serum alanine transaminase	250 U/l	10-45 U/l
Serum aspartate transaminase (AST)	550 U/l	10-50 U/l
Serum γ-glutamyl transpeptidase (GGT)	700 U/l	5-50 U/l

1. **What is the probable diagnosis?**
2. **What is explanation for elevation in the activities of 3 enzymes?**

Ans. :

1. **Alcoholic liver disease.**
2. These 3 enzymes leak into the circulation due to hepatocellular damage.

Clinical Case Study 20.5
(Source : Biochemistry, Page 342)

A 4-year old male child was passing less quantity of urine that was red in colour. He was found to have marked edema around the eyes. The child had the history of repeated upper respiratory tract infections. Microscopic examination of urine showed erythrocytes, leucocytes, epithelial cells and granular casts. The child had almost normal concentrations of blood urea, serum creatinine, serum total proteins and albumin. His 24 hour urine contained 1000 mg protein (normal < 150 mg).

1. **What is the probable diagnosis?**
2. **Is this disorder reversible?**

Ans. :

1. **Acute glomerulonephritis.**
2. In most cases, acute glomerulonephritis is reversible.

Clinical Case Study 20.6
(Source : Biochemistry, Pages 183-185)

A 20-year old man had generalized edema of the body with puffiness of the face in the mornings. His laboratory findings are given

Parameter	Subject	Reference range
Serum total proteins	4.5 g/dl	6-8 g/dl
albumin	1.5 g/dl	3.5-5 g/dl
globulins	3.0 g/dl	2.5-3.5 g/dl
Serum cholesterol	350 mg/dl	150-200 mg/dl
Blood urea	30 mg/dl	15-40 mg/dl
Serum creatinine	1.2 mg/dl	0.5-1.5 mg/dl
Urinary proteins	4 g/day	< 100 mg/day

The serum electrophoresis of the subject showed a sharp and prominent α_2-globulin band.

1. **What is the probable diagnosis? Justify.**
2. **What is the reason for edema?**
3. **What is the dietary advice for patients of this disorder?**

Ans. :

1. **Nephrotic syndrome**, characterized by generalized edema heavy proteinuria and hypoproteinemia.
2. Generalized edema is due to low plasma colloidal osmotic pressure, as a result of reduced albumin concentration.

3. Limit salt intake, reduce saturated fat and cholesterol consuming more water.

For more clinical case studies, refer Chapters 10.3, 10.4

VIVA/Short Questions

1. What are the major functions of liver?

Metabolic, excretory, detoxification, hematological and storage functions.

2. What are the laboratory tests to assess liver function?

Tests based on excretory function (serum bilirubin), serum enzymes (determination of transaminases), metabolic capacity (galactose tolerance) and detoxification (hippuric acid synthesis).

3. What is the normal concentration of serum bilirubin?

0.2-1.0 mg/dl; conjugated bilirubin 0.2-0.8 mg/dl, while unconjugated < 0.2 mg/dl.

4. What do you know about van den Bergh reaction?

van den Bergh reaction is a specific test to identify the increase in serum bilirubin level (< 1.0 mg/dl). The test is said to be direct positive (due to conjugated bilirubin), if the purple colour is obtained within 30 seconds. On the addition of methanol (or alcohol), the unconjugated bilirubin gives colour and this is indirect positive test.

5. Name the serum enzymes employed to assess liver function.

Alanine transaminase, alkaline phosphatase, γ-glutamyltranspeptidase and 5′-nucleotidase.

6. What is jaundice and what are the different types?

Jaundice is characterized by yellow coloration of sclera (of eyes) and skin. This is due to elevated serum bilirubin level, usually beyond 2 mg/dl (normal < 1 mg/dl). The three types of jaundice namely hemolytic, obstructive and hepatic are respectively characterized by elevation in unconjugated bilirubin, conjugated bilirubin and both of these.

7. How is the detoxification function of liver assessed?

Measurement of hippuric acid synthesis is an ideal test for assessing the detoxification function of liver.

8. What is glomerular filtration rate?

Glomerular filtration rate (GFR) refers to the volume of plasma filtered per minute by the kidneys. Normal GFR is 120-125 ml/min.

9. What do you understand by renal threshold substances?

The substances whose excretion in urine is dependent on their blood concentration are referred to as renal threshold substances e.g., glucose, ketone bodies.

10. What is renal threshold of a substance?

The concentration of a given substance in blood (or plasma) beyond which it is excreted into urine e.g., renal threshold for glucose 180 mg/dl; for ketone bodies 3 mg/dl; for calcium 10 mg/dl.

11. What do you understand by tubular maximum?

The maximum capacity of the kidneys to absorb a particular substance per minute e.g., tubular maximum for glucose (TmG) is 350 mg/min.

12. What are the laboratory tests to assess kidney function?

Clearance tests for glomerular function (inulin, creatinine), tests for tubular function (urine concentration or dilution), besides the analysis of blood (urea, creatinine) and urine examination (pH, specific gravity, osmolality).

13. Define clearance.

Clearance refers to the volume of plasma that would be completely cleared of a substance per minute. In other words, clearance of a substance refers to the milliliters of plasma which contains the amount of substance excreted by kidneys per minute.

14. What is creatinine clearance and what is its significance?

Creatinine clearance refers to the volume (ml) of plasma that would be completely cleared of creatinine per minute. Normal 120-145 ml/min. A decrease in creatinine clearance (< 75% normal) indicates impairment in renal function due to renal damage.

15. What do you know about urea clearance?

Urea clearance is the volume (ml) of plasma that would be completely cleared of urea per minute. Normal urea clearance is around 75 ml/min.

16. What are the common laboratory tests to assess renal tubular function?

Measurement of osmolality (normal 500-1,200 milliosmoles/kg) and specific gravity (normal 1.020).

17. What are the components in blood/serum commonly estimated in the laboratory for kidney function?

Serum creatinine (normal 0.5-1.5 mg/dl) and blood urea (normal 15-40 mg/dl).

18. What do you understand by alkaline tide?

Following a meal, there is a slight elevation in the plasma bicarbonate level which is linked to the gastric HCl secretion. This is referred to as alkaline tide.

19. What are the different tests to assess gastric function?

Fractional test meal, pentagastrin stimulation test, augmented histamine test meal and insulin test meal.

21. Water, Electrolyte and Acid-Base Balance

MCQs

1. **The organ most actively and directly involved in water, electrolyte and acid-base balance of the body**
   (A) Brain (B) Liver (C) Kidney (D) Heart.
2. **Water performs the following functions except it**
   (A) Participates in metabolic reactions
   (B) Regulates body's temperature
   (C) Generates energy
   (D) Acts as a vehicle for transport of solutes.
3. **The water content of an adult body**
   (A) 10-30% (B) 30-50%
   (C) 50-70% (D) 70-90%.
4. **The major quantity of body's water is present in**
   (A) Plasma (B) Intracellular fluid
   (C) Extracellular fluid (D) Interstitial fluid.
5. **Ingestion of water is mainly controlled by thirst centre located in**
   (A) Hypothalamus (B) Anterior pituitary
   (C) Posterior pituitary (D) Adrenal cortex.
6. **The endogenous water produced within the body by chemical reactions is referred to as metabolic water. The amount of metabolic water generated is the highest (per gram) for the following foodstuff**
   (A) Carbohydrates (B) Fats
   (C) Proteins (D) Vitamins.
7. **The volume of water filtered by the glomeruli into the renal tubules every day is about**
   (A) 2 l (B) 60 l (C) 120 l (D) 180 l.
8. **The hormone controlling water excretion by the kidney**
   (A) Vasopressin (B) Oxytocin
   (C) Aldosterone (D) Cortisone.
9. **The electrolyte closely associated with the water metabolism in the body**
   (A) Calcium (B) Magnesium
   (C) Sodium (D) Potassium.
10. **The principal extracellular cation**
    (A) Na^+ (B) K^+ (C) Ca^{2+} (D) Mg^{2+}.
11. **The principal intracellular cation**
    (A) Na^+ (B) K^+ (C) Ca^+ (D) Mn^{2+}.
12. **Na^+-K^+ pump is responsible for maintaining**
    (A) High extracellular Na^+ and low intracellular K^+
    (B) High extracellular K^+ and low intracellular Na^+
    (C) Equal concentration of Na^+ and K^+ in the extracellular and intracellular fluids
    (D) None of the above.
13. **The normal osmolality of plasma**
    (A) 85-95 milliosmoles/kg
    (B) 185-195 milliosmoles/kg
    (C) 285-295 milliosmoles/kg
    (D) 385-395 milliosmoles/kg.
14. **The largest contribution (~ 90%) to plasma osmolality is due to the following cation and its associated anions**
    (A) Sodium (B) Potassium
    (C) Magnesium (D) Calcium.
15. **The regulation of electrolyte and water balance is brought about by the hormone(s)**
    (A) Aldosterone
    (B) Antidiuretic hormone
    (C) Renin-angiotensin
    (D) All of the above.
16. **The hormone that increases the Na^+ reabsorption by the renal tubules**
    (A) Antidiurectic hormone
    (B) Aldosterone (C) Cortisol
    (D) Renin.
17. **The hormone that controls the secretion of aldosterone by the adrenal gland**
    (A) Antidiuretic hormone
    (B) Renin (C) Angiotensin
    (D) Renin-angiotensin system.
18. **The characteristic feature(s) of dehydration**
    (A) Increased electrolyte concentration and osmotic pressure of plasma
    (B) Elevated secretion of antidiuretic hormone
    (C) Low blood pressure, sunken eyeballs, confusion and coma
    (D) All of the above.

19. The normal pH of blood
(A) 7.15-7.25 (B) 7.25-7.35
(C) 7.35-7.45 (D) 7.45-7.55.

20. The blood pH compatible to life
(A) 6.4-6.8 (B) 6.8-7.8
(C) 7.8-8.2 (D) 8.2-9.2.

21. The most predominant acid produced in the body
(A) Lactic acid (B) Pyruvic acid
(C) Carbonic acid (D) Hydrochloric acid.

22. A diet rich in animal protein results in more production of
(A) Acids by the body and excretion of acidic urine
(B) Bases by the body and excretion of alkaline urine
(C) Acids by the body and excretion of alkaline urine
(D) Bases by the body and excretion of acidic urine.

23. A vegetarian diet has the following effect on the body
(A) Acidifying effect (B) Alkalizing effect
(C) Neutralizing effect (D) None of the above.

24. The body has developed the following line(s) of defense to maintain blood pH
(A) Blood buffers
(B) Respiratory mechanism
(C) Renal mechanism (D) All of the above.

25. The predominant buffer system of plasma
(A) Bicarbonate buffer (B) Phosphate buffer
(C) Hemoglobin buffer (D) Albumin buffer.

26. The normal plasma bicarbonate concentration
(A) 14-18 mmol/l (B) 18-22 mmol/l
(C) 22-26 mmol/l. (D) 26-30 mmol/l.

27. The ratio of plasma bicarbonate to carbonic acid
(A) 5 : 1 (B) 10 : 1 (C) 20 : 1 (D) 40 : 1.

28. In the normal circumstances, the concentration of the following compound(s) determine(s) the blood pH
(A) Plasma proteins and hemoglobin
(B) Bicarbonate and carbonic acid
(C) Sodium dihydrogen phosphate and disodium hydrogen phosphate
(D) All of the above.

29. The amino acid that is the most effective contributor for protein buffering action
(A) Lysine (B) Arginine
(C) Histidine (D) Proline.

30. The respiratory mechanism for blood pH maintenance primarily acts by regulating the concentration of
(A) Hemoglobin (B) Carbonic acid
(C) Bicarbonate (D) Phosphate.

31. The tissue where carbonic acid and bicarbonate are predominantly generated
(A) Liver (B) Brain
(C) Leucocytes (D) Erythrocytes.

32. The HCO_3^- ions diffuse from RBC to plasma in exchange for
(A) OH^- (B) CO_2 (C) Cl^- (D) $NaHPO_4^-$.

33. A permanent solution to acid-base disturbances is provided by
(A) Bicarbonate buffers
(B) Respiratory mechanism
(C) Renal mechanism (D) Protein buffers.

34. Compared to the pH of blood, urine normal pH is
(A) Acidic (B) Alkaline
(C) Neutral (D) The same.

35. Renal regulation of blood pH occurs by
(A) Excretion of ammonium ions
(B) Excretion of H^+ ions
(C) Reabsorption of bicarbonate
(D) All of the above.

36. The H^+ ions generated by the body are eliminated through
(A) Lungs (B) Intestine (C) Kidneys (D) Liver.

37. The removal of H^+ ions through the kidneys is linked with a simultaneous generation of
(A) OH^- (B) HCO_3^-
(C) H_2CO_3 (D) $NaHPO_4^-$.

38. The normal urine is almost free from
(A) NH_4^+ (B) H^+
(C) HCO_3^- (D) $NaHPO_4^-$.

39. The reabsorption of HCO_3^- by the renal tubules is linked with the excretion of
(A) H^+ (B) Na^+
(C) K^+ (D) All of the above.

40. The ammonium (NH_4^+) ions excreted into the urine are derived from
(A) Glutamate (B) Aspartate
(C) Glutamine (D) Asparagine.

41. The acid load of the body is mostly eleminated in the form of
(A) H^+ (B) NH_4^+ (C) H_2CO_3 (D) HCO_3^-.

42. The fate(s) of the CO_2 generated by the body
(A) Exhaled via lungs
(B) Converted to HCO_3^- by RBC
(C) Converted to HCO_3^- by kidneys
(D) All of the above.

43. Metabolic acidosis is primarily due to a/an
(A) Increase in carbonic acid
(B) Decrease in bicarbonate
(C) Increase in bicarbonate
(D) Decrease in carbonic acid.

44. Respiratory alkalosis is largely due to a/an
(A) Decrease in bicarbonate
(B) Increase in carbonic acid
(C) Increase in bicarbonate
(D) Decrease in carbonic acid.

45. The acid-base disturbances are due to alterations in the concentration of
(A) Bicarbonate (B) Carbonic acid
(C) Bicarbonate and carbonic acid
(D) Hemoglobin.

46. The unmeasured anions in the plasma represent anion gap. This is contributed by the following except
(A) Proteins and organic acids
(B) Phosphate and sulfate
(C) Urate (D) Amino acids.

47. The anion gap in healthy individuals
(A) 0-8 mEq/l (B) 8-18 mEq/l
(C) 18-25 mEq/l (D) 25-32 mEq/l.

48. Metabolic acidosis is commonly seen in uncontrolled
(A) Diabetes insipidus (B) Diabetes mellitus
(C) Atherosclerosis
(D) Nephrotic syndrome.

49. The primary defect in respiratory acidosis
(A) Increased H_2CO_3 (B) Decreased HCO_3^-
(C) Decreased H_2CO_3 (D) Increased HCO_3^-.

50. Metabolic alkalosis is mainly due to
(A) Increased H_2CO_3 (B) Decreased H_2CO_3
(C) Increased HCO_3^- (D) Decreased HCO_3^-.

Answers

1. **(C)** 2. **(C)** 3. **(C)** 4. **(B)** 5. **(A)** 6. **(B)** 7. **(C)** 8. **(A)** 9. **(C)** 10. **(A)** 11. **(B)** 12. **(A)** 13. **(C)** 14. **(A)** 15. **(D)** 16. **(B)** 17. **(D)** 18. **(D)** 19. **(C)** 20. **(B)** 21. **(C)** 22. **(A)** 23. **(B)** 24. **(D)** 25. **(A)** 26. **(C)** 27. **(C)** 28. **(B)** 29. **(C)** 30. **(B)** 31. **(D)** 32. **(C)** 33. **(C)** 34. **(A)** 35. **(D)** 36. **(C)** 37. **(B)** 38. **(C)** 39. **(A)** 40. **(C)** 41. **(B)** 42. **(D)** 43. **(B)** 44. **(D)** 45. **(C)** 46. **(D)** 47. **(B)** 48. **(B)** 49. **(A)** 50. **(C)**.

Clinical Case Study 21.1
(Source : Biochemistry, Page 483)

A 55-year old man was brought to the hospital in a confused and semiconscious state. He had low BP and feeble pulse. His breath had fruity odour. The data of his laboratory investigations

Parameter	*Subject*	*Reference range*
Blood pH	7.1	7.35-7.45
Plasma bicarbonate (HCO_3^-)	12 mmol/l	22-26 mmol/l
Plasma carbonic acid (H_2CO_3)	1.2 mmol/l	1.2 mmol/l
Blood glucose (random)	580 mg/dl	<140 mg/dl
Urine sugar	4+	negative
Urine ketone bodies	3+	negative

1. **What is the probable diagnosis?**
2. **What is basis of this disorder?**
3. **Which compound is responsible for the fruity order of breath?**

Ans. :

1. Metabolic acidosis appropriately regarded as diabetic ketoacidosis.
2. This diabetic patient was not under control. The acidosis was due to the accumulation of ketone bodies.
3. Acetone.

Clinical Case Study 21.2
(Source : Biochemistry, Page 483)

One patient with severe vomiting and hyokalemia was admitted in an intensive care unit. His laboratory data and acid-base status are given below.

Parameter	*Subject*	*Reference range*
Blood pH	7.7	7.35-7.45
Plasma bicarbonate (HCO^-_3)	42 mmol/l	22-26 mmol/l
Plasma carbonic acid (H_2CO_3) (pCO_2 40 mm Hg)	1.2 mmol/l	1.2 mmol/l

1. **What is the probable diagnosis?**
2. **Briefly describe the role of blood buffers in maintanance of bolld pH.**

Ans. :

1. Metabolic alkalosis.
3. Refer Chapter 21, pages 476-477.

Clinical Case Study 21.3
(Source : Biochemistry, Page 483)

A woman had severe pneumonia, and was on treatment. The following are the laboratory data on the acid-base status of this woman admitted in a nursing home.

Parameter	*Subject*	*Reference range*
Blood pH	7.1	7.4
Plasma bicarbonate (HCO_3^-)	24 mmol/l	24-30 mmol/l
Plasma carbonic acid (H_2CO_3) (pCO_2 80 mm Hg)	2.4 mmol/l	1.2 mmol/l

1. **What is the probable diagnosis?**
2. **Why the patient had lower blood pH?**
3. **Carbon dioxide is the central molecule of pH regulation-justify.**

Ans. :

1. Respiratory acidosis.
2. Blood pH is lower due to an increased level of carbonic acid.
3. Refer Chapter 21, page 480.

Clinical Case Study 21.4
(Source : Biochemistry, Page 484)

An adult man with a history of anxiety and severe hyperventilation was admitted in an intensive care unit. The data on the acid-base status of this man are given in the next page.

Parameter	*Subject*	*Reference range*
Blood pH	7.7	7.35-7.45
Plasma bicarbonate (HCO_3^-)	24 mmol/l	22-26 mmol/l
Plasma carbonic acid (H_2CO_3) (pCO_2 20 mm Hg)	0.6 mmol/l	1.2 mmol/l

1. **What is the probable diagnosis?**
2. **What is main reason for increased blood pH?**
3. **Briefly describe the role of respiratory mechanism in the regulation of blood pH?**

Ans. :

1. Respiratory alkalosis.
2. pH of this man is increased due to a reduction in the concentration of carbonic acid.
3. Refer Chapter 21, pages 477-478.

For more clinical case studies, refer 19.5, 23.1, 23.2

VIVA/Short Questions

1. **What are the major functions of water?**

Water is the solvent of life. It provides aqueous medium to the organism, participates in several metabolic reactions, serves as a vehicle for transport of solutes and is closely associated with regulation of body temperature.

2. **How is water distributed in the body?**

A 70 kg normal man contains about 42 litres of water. This is distributed in intracellular (28 l) and extracellular (14 l) compartments, respectively known as intracellular fluid (ICF) and extracellular fluid (ECF).

3. **What do you understand by metabolic water?**

The endogenously produced water in the body due to oxidation of foodstuffs. About 125 ml of water is generated for 1,000 Cal consumed.

4. **How is consumption of exogenous water controlled?**

Thirst centre located in the hypothalamus mainly controls water intake.

5. **How is water lost from the body?**

Urine is the major route for water output (1-2 l/ day) from the body. Some amount of water is also lost through skin (450 ml/day), lungs (400 ml/day) and feces (150 ml/day).

6. **How does hormonal regulation of urine formation occur?**

Antidiuretic hormone (ADH, vasopressin) tightly controls the water excretion by the kidney. This is evident from the fact that of the huge volume of about 180 litres of water filtered by the glomeruli into the renal tubules everyday, just 1-2 litres is excreted as urine, while the rest is reabsorbed. An increase in plasma osmolality promotes ADH secretion which causes increased water reabsorption from the renal tubules.

7. **Name the disorder caused by the deficiency of antidiuretic hormone.**

Diabetes insipidus.

8. **What is the formula for the conversion of mg/l to mEq/l?**

$$\text{mEq/l} = \frac{\text{mg per litre} \times \text{valency}}{\text{atomic weight}}$$

9. **How are electrolytes distributed in the body?**

Na^+ is the principal extracellular cation, while K^+ is the intracellular cation. As regards anions, Cl^- and HCO_3^- predominantly occur in extracellular fluids while $HPO_4^=$, proteins and organic acids are found in the intracellular fluids.

10. What do you understand by osmolarity and osmolality?

Osmolarity refers to the number of moles (or millimoles) per liter of solution while osmolality means the number of moles (or millimoles) per kg of solvent.

11. What is the normal plasma osmolality and how can it be computed from the estimated plasma Na^+ concentration?

The normal osmolility of plasma is in the range of 285-295 milliosmoles/kg.

$$\text{Plasma osmolality (mmol/kg)} = 2 \times \text{plasma } Na^+ \text{ (mmol/l)}.$$

12. Name the hormones that regulate electrolyte and water balance in the body.

Aldosterone, antidiuretic hormone and renin-angiotensin.

13. How does aldosterone influence electrolyte metabolism?

Aldosterone increases Na^+ reabsorption by renal tubules at the expense of K^+ and H^+ ions. The net effect is retention of Na^+ in the body.

14. What is the normal pH of blood?

Normal blood pH 7.35-7.45.

15. Name the important acids produced by the body.

Carbonic acid (most predominant), lactic acid, sulfuric acid and phosphoric acid.

16. How are the acids produced by the body?

Carbonic acid is formed from the metabolic product CO_2; lactic acid is produced by anaerobic metabolism; sulfuric acid generated from sulfur containing amino acids.

17. What are the different methods adopted by the body to regulate blood pH?

Three lines of defense to regulate acid-base balance and maintain the blood pH around 7.4. These are blood buffers, respiratory mechanism and renal mechanism.

18. Name the buffer systems involved in maintaining blood pH.

Bicarbonate buffer, phosphate buffer and protein buffers.

19. What is Henderson-Hasselbalch equation?

$$pH = pKa + \log \frac{(\text{base})}{(\text{acid})}$$

20. Which is the most predominant buffer that maintains blood pH?

Bicarbonate (HCO_3^-) and carbonic acid (H_2CO_3) buffer system. Normal plasma HCO_3^- = 24 mmol/l; H_2CO_3 = 1.2 mmol/l. The ratio of bicarbonate to carbonic acid is 20 : 1.

21. What do you understand alkali reserve?

At a pH of blood 7.4, the ratio of bicarbonate to carbonic acid is 20 : 1. Thus, the bicarbonate concentration is much higher (20 times) than carbonic acid in blood. This is referred to as alkali reserve, and is responsible for the effective buffering of H^+ ions, generated by the body.

22. What is the role of respiratory system in blood pH regulation?

Respiratory system provides a rapid mechanism for the maintenance of acid-base balance. Large volume of CO_2 produced by the cellular metabolic activity is eliminated from the body in the expired air via lungs.

23. How is bicarbonate generated by RBC?

The plasma bicarbonate diffuses into the RBC along with concentration gradient where it combines with water to form H_2CO_3 (catalysed by carbonic anhydrase). In the RBC, H_2CO_3 dissociates to produce H^+ and HCO_3^-.

24. What is chloride shift?

HCO_3^- diffuses into plasma along with the concentration gradient, in exchange for Cl^- ions to maintain electrical neutrality. This is referred to as chloride shift.

25. How do kidneys regulate blood pH?

The renal mechanism of pH regulation involves excretion of H^+ ions, titratable acid and ammonium ions, besides the reabsorption of bicarbonate.

26. What is titratable acidity?

The acid excreted into urine by kidneys. This can be estimated by titrating urine back to the normal pH of blood (7.4).

27. Carbon dioxide is the central molecule of pH regulation—justify.

Erythrocytes, lungs and kidneys coordinate to handle CO_2 and maintain blood pH at 7.4. In essense, CO_2 produced by the aerobic metabolism may be exhaled via lungs, or converted to HCO_3^- by erythrocytes and kidneys to add up to the alkali reserve of the body.

28. What is the blood pH compatible for existence of life?

pH 6.8-7.8.

29. What is acidosis?

A decline in blood pH. It is mainly of two types — metabolic acidosis due to a decrease in bicarbonate; respiratory acidosis due to an increase in carbonic acid.

30. What do you know about alkalosis?

A rise in blood pH. Metabolic alkalosis is caused by an increase in bicarbonate while respiratory alkalosis is due to a decrease in carbonic acid.

31. What is anion gap?

The unmeasured (in the laboratory) anions in the plasma. The difference between the total concentration of measured cations (Na^+ and K^+) and that of measured anions (Cl^- and HCO_3^-)

32. What do you know about metabolic acidosis?

A fall in the blood pH, primarily due to a reduction in bicarbonate concentration.

33. How is metabolic alkalosis compensated?

Primarily due to an increase in HCO_3^- concentration.

22. Tissue Proteins and Body Fluids

MCQs

1. The amino acid that contributes to one-third of the total number of amino acids in collagen

(A) Serine (B) Glycine
(C) Proline (D) Lysine.

2. The following amino acids confer rigidity to collagen molecule

(A) Proline and hydroxyproline
(B) Glycine and tryptophan
(C) Phenylalanine and serine
(D) Cysteine and cystine.

3. The functional unit of muscle

(A) Tropomyosin (B) Troponin
(C) Sarcomere (D) Myosin.

4. The most abundant protein in milk

(A) Albumin (B) Lactalbumin
(C) Lactoglobulin (D) Casein.

5. The fetal ling maturity is evaluated by the following, in amniotic fluid

(A) Glucose
(B) Lecithin sphingomyelin ratio
(C) α-Fetoprotein (D) Creatinine.

6. A man aged 65 years had confusion, hallucinations, memory loss etc. He was diagnosed as a patient of Alzheimer's disease, caused by aggregation of the following protein in the brain.

(A) Histones (B) β-amyloid
(C) Albumin (D) Globulins

Answers

1. **(B)** 2. **(A)** 3. **(C)** 4. **(D)** 5. **(B)** 6. **(B)**

Clinical Case Study 22.1 (Source : Biochemistry, Page 496)

A 6-year old boy had high temperature with sweats during sleep, marked tachycardia and loss of weight. He also had the complaints of headache, neck stiffness and vomiting. The liver was found to be slightly enlarged. The following are the laboratory findings of this boy.

Parameter	*Subject*	*Reference range*
Cerebrospinal fluid (CSF)		
Glucose	30 mg/dl	50-75 mg/dl
Proteins	100 mg/dl	15-40 mg/dl
Chloride	105 mEq/l	120-130 mEq/l
Blood glucose	100 mg/dl	70-100 mg/dl
(simultaneously collected)		

1. What is the probable diagnosis with justification?

2. Write briefly on the importance of CSF in the diagnosis.

Ans. :

1. A decrease in the CSF glucose and chloride concentrations along with an increase in protein level indicates that the boy was a victim of **tuberculous meningitis**.
2. Refer Chapter 22, page 496.

For related clinical case study, refer 7.3

VIVA/Short Questions

1. What is collagen?

The most abundant protein in mammals, comprising one-third of the total body protein.

2. What are the major functions of collagen?

Gives strength, support and shape to the tissues, alignment of cells.

3. How is collagen related to aging?

Collagen production declines with increasing age. This results in wrinkles and sagging skin, besides muscle weakness.

4. What do you understand by collagenopathies?

Collagenopathies are inherited diseases due to gene mutations, linked with collagen formation. e.g., Ehlers-Danlos syndrome (EDS), Osteogenesis imperfecta.

5. Name the vitamin linked to improper formation of collagen (due to defect in post-translational modification).

Vitamin C (ascorbic acid).

6. What is Marfan syndrome?

A genetic disorder due to a mutation in the gene for fibrillin. Patients of Marfan syndrome are tall, and have long digits.

7. Name the structural proteins found in hair, skin, nails and horns?

Keratins

8. What do you understand by sarcopenia?

Loss of skeletal muscle mass with age.

9. Name the major disease linked to protein-misfolding.

Prion diseases e.g., kuru, mad cow disease

amyloidosis e.g., Alzheimer's disease.

10. Why the colour of milk is white?

Due to the dispersion of calcium caseinate micelles.

11. Name some nutrients deficient or limiting in milk

Vitamin C, iron, copper.

12. How is the biochemical profile of CSF altered in tuberculosis meningitis?

The total cell count and protein content are increased while glucose concentration is reduced.

13. What is the diagnostic importance of amniotic fluid?

Assessment of fetal maturity : Fetal maturity can be assessed by cytological staining of fat cells, and estimation of creatinine concentration (> 1.6 mg/dl indicates fetal maturity).

Lung maturity : The fetal lung maturity is evaluated by measuring lecithin–sphingomyelin (L/S) ratio. A L/S ratio of 2 : 1 or more indicates lung maturity. If L/S ratio is less than 1.2 : 1, it is better to delay the induced delivery until the lung has become more mature.

23. Nutrition

MCQs

1. Calorie is the amount of heat required to rise the temperature of one gram of water by

(A) 1°C (B) 2°C (C) 3°C (D) 4°C.

2. One calorie is equivalent to (joules)

(A) 1.128 (B) 2.128 (C) 3.128 (D) 4.128.

3. The calorific value of proteins (in Cal/g)

(A) 2 (B) 4 (C) 6 (D) 9.

4. The respiratory quotient for fats

(A) 0.4 (B) 0.7 (C) 1.0 (D) 1.3.

5. The body process that consumes a major part of the energy under basal conditions

(A) Conduction of nerve impulse

(B) Reabsorption by renal tubules

(C) Digestion and absorption of foodstuffs

(D) Na^+-K^+ pump.

6. The standard calorific value for 1 litre of O_2 consumed

(A) 2.825 Cal (B) 3.825 Cal

(C) 4.825 Cal (D) 5.825 Cal.

7. The normal BMR for an adult man (Cal/sq. m/hr)

(A) 30-35 (B) 35-40 (C) 40-45 (D) 45-50.

8. The following statements regarding the factors affecting BMR are true except

(A) The body surface area is directly proportional to BMR

(B) Men have lower (about 5%) BMR than women

(C) Thyroid hormones stimulate BMR

(D) BMR is increased in persons with regular exercise.

9. The endocrine gland largely responsible for the regulation of BMR

(A) Pituitary (B) Pancreas

(C) Adrenal (D) Thyroid.

10. The extra heat produced by the body over and above the calculated caloric value is referred to as

(A) Basal metabolic rate

(B) Energy sparing action

(C) Specific dynamic action

(D) Respiratory quotient.

11. The specific dynamic action for protein, fat and carbohydrate respectively

(A) 30%, 13% and 5%

(B) 13%, 30% and 5%

(C) 5%, 30% and 13%

(D) 30%, 5% and 13%.

12. For a mixed diet, the specific dynamic action is around

(A) 5% (B) 10% (C) 25% (D) 30%.

13. Based on the specific dynamic action, the best source of energy

(A) Carbohydrate (B) Protein

(C) Fat (D) Vitamins.

14. The specific dynamic action of proteins is attributed to
(A) Synthesis of urea
(B) Biosynthesis of proteins
(C) Deamination of amino acids
(D) All of the above.

15. One of the following physical activities requires the highest energy (Cal/hr)
(A) Running (B) Cycling
(C) Swimming (D) Walking upstairs.

16. In an individual with light work, the energy expenditure towards BMR, physical activity and specific dynamic action respectively
(A) 60%, 10% and 30%
(B) 60%, 30% and 10%
(C) 30%, 60% and 10%
(D) 10%, 60% and 30%.

17. The dietary nutrients that are not absolutely essential to the body
(A) Carbohydrates (B) Proteins
(C) Lipids (D) Vitamins.

18. The chief source of energy to the body under normal circumstances
(A) Carbohydrates (B) Proteins
(C) Lipids
(D) Essential amino acids.

19. The most abundant dietary carbohydrate supplied to the body
(A) Sucrose (B) Glucose
(C) Starch (D) Cellulose.

20. Excessive inake of the following carbohydrate is associated with increased plasma lipid levels
(A) Sucrose (B) Glucose
(C) Starch (D) Cellulose.

21. Carbohydrates can spare the following nutrients from their wasteful expenditure to meet the energy demands of the body
(A) Fats (B) Proteins
(C) Vitamins
(D) Essential fatty acid.

22. The organ absolutely dependent on glucose for energy demands
(A) Skeletal muscle (B) Heart
(C) Liver (D) Brain.

23. The term dietary fiber refers to the complex carbohydrates that are
(A) Digested and absorbed
(B) Digested but not absorbed
(C) Non-digestible
(D) Not digested but absorbed.

24. The following carbohydrates are regarded as dietary fiber except
(A) Cellulose and hemicellulose
(B) Pectin and lignin
(C) Gums and mucilages
(D) Starch and glycogen.

25. The following are the beneficial effects of dietary fiber except it
(A) Prevents constipation and eliminates bacterial toxins
(B) Improves glucose tolerance and reduces serum cholesterol
(C) Decreases gastrointestinal cancers
(D) Decreases the intestinal absorption of calcium and phosphate.

26. The dietary source of lipid that largely contributes to the energy needs of the body
(A) Phospholipids (B) Fatty acids
(C) Triacylglycerols (D) Cholesterol.

27. The following are the nutritional functions of lipids except that they do not
(A) Supply concentrated source of energy
(B) Provide essential fatty acids
(C) Supply fat soluble vitamins
(D) Provide nitrogen.

28. In a strict sense, the following two are essential fatty acids
(A) Oleic and linoleic acid
(B) Linoleic and linolenic acid
(C) Linolenic and arachidonic acid
(D) Linoleic and arachidonic acid.

29. Essential fatty acids primarily perform the following functions except they
(A) Participate in the transport and utilization of cholesterol
(B) Prevent fat accumulation in liver
(C) Are required for the synthesis of prostaglandins
(D) Provide energy to the body.

30. Phrynoderma, a disorder characterized by scaly dermatitis, impairment in growth and reproduction, is due to the deficiency of
(A) Essential amino acids
(B) Essential fatty acids
(C) Dietary fiber (D) Vitamin B_6.

31. The rich sources of polyunsaturated fatty acids include the following except
(A) Sunflower oil (B) Soyabean oil
(C) Fish oil (D) Ghee.

32. High consumption of polyunsaturated fatty acids should be accompanied by one of the following vitamins to protect the body from harmful effects
(A) Vitamin B_1 (B) Vitamin K
(C) Vitamin D (C) Vitamin E.

33. The amino acid cysteine can spare the requirement of
(A) Phenylalanine (B) Methionine
(C) Arginine (D) Lysine.

34. The nutrient that is almost an exclusive source of nitrogen to the body
(A) Carbohydrate (B) Lipid
(C) Nucleic acid (D) Protein.

35. The nitrogen from the body is lost via
(A) Feces (B) Urea in urine
(C) Ammonia in urine
(D) All of the above.

36. A positive nitrogen balance means that the nitrogen intake is
(A) Higher than the output
(B) Lower than the output
(C) Equal to that of the output
(D) None of the above.

37. A positive nitrogen balance is observed in the following conditions except in
(A) Growing children
(B) Pregnant women
(C) Recovery after serious illness
(D) Kwashiorkor.

38. One of the following hormone(s) causes negative nitrogen balance
(A) Growth hormone (B) Insulin
(C) Epinephrine (D) Cortisol.

39. The gain in the body weight of rats per gram protein consumed represents
(A) Biological value of proteins
(B) Protein efficiency ratio
(C) Net protein utilization
(D) Chemical score.

40. Biological value of a protein represents the
(A) Percentage of nitrogen retained by the body
(B) Gain in body weight per gram protein consumed
(C) Essential amino acid composition of proteins
(D) None of the above.

41. The protein with highest biological value
(A) Meat protein (B) Fish protein
(C) Milk protein (D) Egg protein.

42. Rice and wheat proteins are limiting in the essential amino acids
(A) Lysine and methionine
(B) Lysine and threonine
(C) Threonine and phenylalanine
(D) Leucine and threonine.

43. The deficiency of the following essential nutrients is uncommon
(A) Essential fatty acids
(B) Essential amino acids
(C) Calcium and phosphorus
(D) Vitamins C and D.

44. The requirement of protein by an adult (g/kg body weight/day)
(A) 0.4–0.6 (B) 0.6–0.8
(C) 0.8–1.0 (D) 1.0–1.2.

45. The recommended dietary allowance (RDA) for men, in general is higher (about 20%) than for women. The following nutrient, however, is an exception
(A) Vitamin C
(B) Calcium
(C) Essential fatty acids
(D) Iron.

46. The term kwashiorkor literally means
(A) Sickness of the deposed child
(B) Sickness of the new born
(C) Sickness of the preschool child
(D) Presence of moon face.

47. Kwashiorkor is primarily due to insufficient intake of
(A) Carbohydrates (B) Fats
(C) Proteins (D) Vitamins.

48. Kwashiorkor is associated with the following character(s)
(A) Stunted growth and anemia
(B) Fatty liver and diarrhea
(C) Decreased plasma albumin concentration
(D) All of the above.

49. Marasmus is characterized by the following except it
(A) Is predominantly due to the deficiency of calories
(B) Occurs in children under 1 year age
(C) Is associated with muscle wasting and anemia
(D) Shows edema and decreased plasma albumin concentration.

50. The term non-exercise activity thermogenesis (NEAT) refers to energy used for

(A) Walking to work

(B) Climbing steps and fidgeting

(C) Performing kitchen and yard work

(D) All of the above.

51. High fructose corn syrup (HFCS), used as a substitute for sucrose in beverages, is produced from

(A) Inulin (B) Fructose

(C) Glucose (D) Sucrose.

52. The measure of time course post-prandial glucose concentration from a graph represents

(A) Glucose tolerance test

(B) Glycemic index

(C) Intestinal glucose absorption

(D) All of the above.

53. One of the following fatty acids is widely used in infant food formulas to promote brain development

(A) Arachidonic acid (B) Linoleic acid

(C) Decosahexaenoic acid

(D) Prostanoic acid

54. The following fatty acids are used in food industry due to long shelf life

(A) Trans fatty acids (B) Cis fatty acids

(C) Prostanoic acids (D) Prostaglandins.

55. The distinction flavour of browned foods (cookies, biscuits, bread) is due the following chemical reaction

(A) Biuret reaction (B) Maillard reaction

(C) Salkowski reaction (D) Haworth reaction

56. The following components contribute to metabolic syndrome

(A) Insulin resistance (B) Dyslipidemia

(C) Hypertension (D) All of the above

57. Atkins diet is composed of

(A) High fat, protein and low carbohydrate diet

(B) High carbohydrate, protein and low fat diet

(C) High protein, and low fat and carbohydrate diet

(D) None of the above

Answers

1. **(A)** 2. **(D)** 3. **(B)** 4. **(B)** 5. **(D)** 6. **(C)** 7. **(B)** 8. **(B)** 9. **(D)** 10. **(C)** 11. **(A)** 12. **(B)** 13. **(C)** 14. **(D)** 15. **(D)** 16. **(B)** 17. **(A)** 18. **(A)** 19. **(C)** 20. **(A)** 21. **(B)** 22. **(D)** 23. **(C)** 24. **(D)** 25. **(D)** 26. **(C)** 27. **(D)** 28. **(B)** 29. **(D)** 30. **(B)** 31. **(D)** 32. **(D)** 33. **(B)** 34. **(D)** 35. **(D)** 36. **(A)** 37. **(D)** 38. **(D)** 39. **(B)** 40. **(A)** 41. **(D)** 42. **(B)** 43. **(B)** 44. **(C)** 45. **(D)** 46. **(C)** 47. **(C)** 48. **(D)** 49. **(D)** 50. **(D)** 51. **(C)** 52. **(B)** 53. **(C)** 54. **(A)** 55. **(B)** 56. **(D)** 57. **(A)**.

Clinical Case Study 23.1
(Source : Biochemistry, Page 515)

A 3-year old female child had stunted growth, edema, apathy and moon-face. She also had frequent respiratory infections and diarrhea. The mother informed the physician that the child was being given dilute buffalo milk and a small quantities of rice with ghee and dhal. The laboratory data of the child

Parameter	*Subject*	*Reference range*
Hemoglobin	7 g/dl	12-14 g/dl
Serum proteins (total)	4 g/dl	6-8 g/dl
Serum albumin	2 g/dl	3-4.5 g/dl
Serum potassium	2.8 mEq/l	3.5-4.5 mEq/1

1. **What is the probable diagnosis?**
2. **How do you explain the presence of edema?**
3. **What is the probable cause of reduced serum potassium?**

Ans. :

1. The child was suffering from **kwashiorkor**.
2. Edema occurred due to lack of adequate serum proteins to maintain water distribution between blood and tissues.
3. Deficiency of serum K^+ was observed due to diarrhea.

Clinical Case Study 23.2
(Source : Biochemistry, Page 515)

A one year old male baby had growth retardation, muscle wasting (emaciation), loose folds of skin. The mother informed the physician that she had been giving the child rice gruel, along with breast milk. The laboratory data of the child

Parameter	*Subject*	*Reference range*
Hemoglobin	7 g/dl	12-14 g/dl
Serum proteins (total)	6 g/dl	6-8 g/dl
Serum albumin	3 g/dl	3-4.5 g/dl
Serum potassium	3.5 mEq/l	3.5-5 mEq/1

1. **What is the probable diagnosis?**

2. **What is the most predominant cause of this disorder?**
3. **How is it distinguished from kwashiorkor?**

Ans. :

1. The child was suffering from **marasmus**.
2. Due to the deficiency of calories.
3. It can be distinguished from kwashiorkor by lack of edema and almost unaltered serum albumin level.

For more clinical case studies, refer 7.1, 7.8, 16.1, 18.1, 19.1, 19.2

VIVA/Short Questions

1. What is calorie?

Calorie is the unit of heat. One calorie represents the amount of heat required to rise the temperature of one gram of water by 1·C (i.e., from 15· to 16·C).

2. How are the calories usually expressed?

As kilocalories (1,000 calorie) which is represented by Kcal or Cal (with capital 'C').

3. What is joule?

The joule is also a unit of energy, used in some countries. The relationship between calories and joules (J) is

1 Cal (Kcal) = 4.128 KJ.

4. What do you know about the calorie value of foods?

The energy values of carbohydrates, fats and proteins (when utilized in the body), respectively are 4, 9 and 4 Cal/g.

5. Name some nutrients that do not have any calorific value.

Vitamins and minerals.

6. What is respiratory quotient?

Respiratory quotient (R.Q.) is the ratio of volume of CO_2 produced to the volume of O_2 utilized in the oxidation of foodstuffs. R.Q. for carbohydrates is about 1; for fats 0.7 and for proteins 0.8; for a mixed diet, around 0.8.

7. What are the major processes in the body that require energy?

Basal metabolic rate, specific dynamic action and physical acitivity. Besides these, additional energy supply is needed during growth, pregnancy and lactation.

8. What is basal metabolic rate (BMR)?

The minimum amount of energy required by the body to maintain life at complete physical and mental rest in the post- absorptive state (i.e., 12 hr after last meal).

9. What are the units of BMR?

BMR is expressed as calories per square meter of body surface area per hour, i.e., Cal/sq. m/hr.

10. What are the normal values of BMR?

For adult man 35-38 Cal/sq. m/hr; for adult woman 32-35 Cal/sq. m/hr (Some authors continue to express BMR as Cal/day i.e., for adult man—1,600 Cal/day; for adult woman—1,400 Cal/day).

11. Name the factors that influence BMR?

Surface area, sex, age, physical acitivity, hormones, environment, starvation, disease states etc.

12. What is specific dynamic action?

The extra heat produced by the body, over and above the calorie value, when a given foodstuff is metabolised by the body. It is also known as calorigenic action or thermogenic action.

13. What are the SDA values for different foods?

SDA for protein, fat and carbohydrate respectively are 30%, 13% and 5%; for a mixed diet around 10%.

14. How is the energy spent by an individual with light work?

About 60% calories towards BMR, 30% for physical activity and 10% to take care of SDA.

15. Name some carbohydrates utilized by the body.

Starch, glycogen, sucrose, lactose, glucose and fructose. Among these, starch is the most abundant.

16. What is dietary fiber?

The complex carbohydrates that are not digested by the human enzymes are collectively referred to as dietary fiber. These include cellulose, hemicellulose, pectin, lignin, gums and mucilages.

17. What do you know about the nutritional importance of dietary fiber?

Fiber prevents constipation, eliminates bacterial toxins, decreases gastrointestinal cancers, improves glucose tolerance, reduces plasma cholesterol and provides satiety value.

18. What are the adverse affects of dietary fiber?

The dietary fiber in large quantities adversely affects protein digestion and absorption, besides decreasing the absorption of certain minerals.

19. What are essential fatty acids?

Linoleic and linolenic acids are essential to humans. Some authors consider arachidonic acid also.

20. What are the deficiency manifestations of essential fatty acids?

Scaly dermatitis on the posterior and lateral parts of limbs and buttocks. This condition is referred to as phrynoderma or toad skin.

21. Name the rich sources of essential fatty acids.

Vegetable oils and fish oils. The rich vegetable sources include sunflower oil, cotton seed oil, corn oil and soyabean oil.

22. What is the requirement of dietary fat?

The recommended fat intake is around 20-30% of the daily calorie requirement containing about 50% of PUFA.

23. Name the essential amino acids.

There are 8 essential amino acids — valine, isoleucine, leucine, lysine, methionine, phenylalanine, tryptophan and threonine.

24. What do you understand by nitrogen balance and nitrogen equilibrium?

Nitrogen balance represents the protein (16% of which is nitrogen) utilization and its loss from the body. A normal healthy adult is in a nitrogen equilibrium, since the daily nitrogen intake (I) is equal to the loss through (U) feces (F) and sweat (S)

Thus, $I = U + F + S$.

25. What is positive nitrogen balance?

The nitrogen intake is higher than the output. Some amount of nitrogen is retained in the body causing a net increase in the body proteins. Positive nitrogen balance is observed in growing children, and pregnant women.

26. When does negative nitrogen balance occur?

Represents a situation wherein nitrogen output is higher than the input. This is mostly due to inadequate dietary intake of proteins, commonly observed in children suffering from kwashiorkor or marasmus.

27. How is the nutritive value of proteins assessed?

By laboratory methods protein efficiency ratio, biological value, net protein utilization and chemical score.

28. What is protein efficiency ratio?

Represents the gain in body weight per gram protein ingested.

29. What do you understand by biological value of protein?

The percentage of absorbed nitrogen retained in the body.

30. What is chemical score?

The ratio between the quantity of the most limiting essential amino acid in the test protein to the quantity of the same amino acid in the egg proteins.

31. What do you understand by mutual supplementation of proteins (or amino acids)?

By consuming a mixed diet with different foods simultaneously, the deficiency of one or more essential amino acids in one food item is supplemented by their presence in the other.

32. Why the dietary deficiency of essential amino acids is the most uncommon?

Due to the mutual supplementation of proteins by consuming a mixed diet essential amino acid deficiency is rarely observed.

33. What is the daily requirement of protein by an adult?

0.8-1.0 g protein/kg body weight/day.

34. What do you understand by the recommended dietary allowance (RDA)?

The RDA represents the quantities of the nutrients to be provided in the diet daily for maintaining good health and physical efficiency of the body.

35. What are the factors that influence RDA?

In general, the RDA for men is about 20% higher than that for women. Iron, however, is an exception, since the requirement is greater in menstruating women. Further, additional requirements (20-30% above normal) are needed for growing children, pregnant, and lactating women.

36. What is a balanced diet?

A balanced diet represents a diet that contains different types of foods, possessing the nutrients — carbohydrates, fats, proteins, vitamins and minerals — in a proportion to meet the requirements of the body.

37. What are ingradients of Indian balanced diet?

The Indian balanced diet is composed of cereals (rice, wheat, jowar), pulses, vegetables, roots and tubers, fruits, milk and milk products, fats and oils, sugar and groundnuts. Meat, fish and eggs are present in the non-vegetarian diets.

38. What are the extreme forms of protein energy malnutrition in pre-school children?

Kwashiorkor and marasmus.

39. What is kwashiorkor?

Due to insufficient intake of proteins and is predominantly found in children between 1-5 years of age. The clinical manifestations include stunted growth, edema, anemia, diarrhea, apathy and moon face. Kwashiorkor is associated with decreased plasma albumin concentration.

40. What do you know about marasmus?

Due to the deficiency of calories and is usually observed in children under one year age. The manifestations of marasmus include growth retardation, muscle wasting (emaciation) anemia and weakness.

41. What are the major distinguishing features between marasmus and kwashiorkor?

A marasmic child does not show edema or decreased concentration of plasma albumin. This is in contrast to kwashiorkor.

MOLECULAR BIOLOGY AND BIOTECHNOLOGY V

24. DNA-Replication, Recombination, and Repair

MCQs

1. **The macromolecule regarded as the reserve bank of genetic information**
   (A) Protein (B) DNA
   (C) RNA (D) Polysaccharide.
2. **The biological information flows from DNA to RNA and from RNA to**
   (A) Lipids (B) Carbohydrates
   (C) Proteins (D) Nucleotides.
3. **The total genetic information contained in a DNA is referred to as**
   (A) Gene (B) Genome
   (C) Okazaki piece (D) Ribozyme.
4. **Replication is characterized by the following except it is**
   (A) Semiconservative (B) Bidirectional
   (C) Semidiscontinuous
   (D) Dependent on DNA primer.
5. **The replication of DNA occurs in the direction of**
   (A) $5 \rightarrow 3$ (B) $5 \rightarrow 3$ (C) $3 \rightarrow 5$ (D) $3 \rightarrow 5$.
6. **The lagging strand is characterized by**
   (A) A constant supply of RNA primers
   (B) Discontinuous synthesis of new strand
   (C) Replication of DNA in $5' \rightarrow 3'$ direction
   (D) All of the above.
7. **The separation of two DNA strands for replication is brought about by**
   (A) DNA polymerase III
   (B) DNA helicases (C) DNA polymerase I
   (D) DNA ligase.
8. **The presence of the following four deoxyribonucleoside triphosphates is essential for replication**
   (A) dATP, dUTP, dCTP, dTTP
   (B) dATP, dGTP, dUTP, dTTP
   (C) dATP, dGTP, dCTP, dTTP
   (D) dATP, dGTP, dCTP, dUTP.
9. **The synthesis of new DNA (replication) in prokaryotes is catalysed by**
   (A) DNA polymerase I (B) DNA polymerase II
   (C) DNA polymerase III
   (D) DNA topoisomerase.
10. **The term Okazaki fragments refers to the**
    (A) Discontinuously synthesized DNA pieces on the lagging strand
    (B) Discontinuously synthesized DNA pieces on the leading strand
    (C) DNA fragments formed by the action of DNA ligase
    (D) None of the above.
11. **The proof-reading activity of the newly synthesized DNA is present on the enzyme**
    (A) DNA helicase
    (B) DNA polymerase I
    (C) DNA polymerase II
    (D) DNA polymerase III.
12. **Ihe problem of supercoils during DNA replication is overcome by a group of enzymes called**
    (A) DNA topoisomerases
    (B) DNA ligases (C) DNA polymerases
    (D) DNA helicases.
13. **The enzyme responsible for the replication of mitochondrial DNA**
    (A) DNA polymerase α
    (B) DNA polymerase β
    (C) DNA polymerase γ
    (D) DNA polymerase δ.
14. **The following disorder(s) is/are attributed to defects in DNA repair**
    (A) Xeroderma pigmentosum
    (B) Fanconi's syndrome
    (C) Cockayne's syndrome
    (D) All of the above.

Answers

1. **(B)** 2. **(C)** 3. **(B)** 4. **(D)** 5. **(B)** 6. **(D)** 7. **(B)** 8. **(C)** 9. **(C)** 10. **(A)** 11. **(D)** 12. **(A)** 13. **(C)** 14. **(D)**.

For related clinical case study, refer 1.1

VIVA/Short Questions

1. **What do you understand by the term genome?**

The total genetic information contained in the DNA is referred to as genome.

2. **What is replication?**

It is a process in which DNA copies itself to produce identical daughter molecules of DNA.

3. **'Replication is semiconservative'—explain.**

Each one of the newly synthesized DNA has one-half of the parental DNA (one strand from the original) and one-half of new DNA strand. The replication is semiconservative since half of the original DNA is conserved in the daughter DNA.

4. Name the protein responsible for the initiation of replication.

Protein dnaA (with 20-50 monomers).

5. In which direction DNA synthesis (replication) occurs?

5′ → 3′ direction.

6. What do you understand by a leading strand and a lagging strand of DNA?

On leading strand is the DNA synthesis is continuous while lagging strand DNA synthesis is discontinuous.

7. What are DNA helicases?

The enzymes that bin DNA strands at the replication fork. Their function is comparable to a zip opener as they separate the two strands of DNA.

8. How are the two strands of DNA kept apart to allow the synthesis of new DNA strands?

Single stranded DNA binding (SSB) proteins bind to the DNA strands (separated by helicases) and keep them apart to allow new DNA synthesis.

9. What do you know about Okazaki pieces?

The small fragments of the discontinuously synthesized DNA are called Okazaki pieces. They are produced on the lagging strand of the parent DNA, and are joined together by DNA ligase.

10. What are the functions of DNA polymerase III?

It catalyses the synthesis of new strand of DNA. In addition, this enzyme possesses proof-reading activity as it checks the incoming nucleotides and allows only the correctly matched bases.

11. What is an RNA primer?

For the synthesis of a new DNA strand, a short fragment of RNA (about 5-50 nucleotides) is required as a primer.

12. What do you know about supercoils?

As the double helix of DNA separates from one side and replication proceeds, supercoils are formed at the other side. Supercoils interfere in DNA replication. This problem is overcome through the participation of DNA topoisomerases.

13. What are the common types of DNA damages?

Single base alterations, two base alterations, chain breaks and cross - links.

14. What is xeroderma pigmentosum?

It is a rare autosomal recessive genetic disease due to a defect in the repair of the DNA damage, caused by ultraviolet rays.

25. Transcription and Translation

MCQs

1. The functional unit of DNA

(A) Genome (B) Gene
(C) Nucleotide (D) Chromosome.

2. Transcription is characterized by

(A) One strand of DNA serving as template
(B) Synthesis of RNA on a selected region of DNA
(C) Participation of RNA polymerase
(D) All of the above.

3. The RNA polymerase of *E.coli* is a complex holoenzyme composed of the following 5 polypeptide subunits

(A) 2α,2β and one sigma factor
(B) 2α,2β and one rho factor
(C) 2α,1β,1β′ and one sigma factor
(D) 2α,1β,1β and one rho factor.

4. Pribnow box consists of the following 6 nucleotide bases

(A) TATAAT (B) TATATA
(C) TTAAAT (D) TAATAT.

5. RNA polymerase utilizes the following ribonucleotide triphosphates for the formation of RNA

(A) ATP, GTP, CTP and TTP
(B) ATP, GTP, CTP and UTP
(C) ATP, CTP, UTP and UTP
(D) None of the above.

6. During the course of transcription (or replication) for the addition of each nucleotide to the growing chain, the following moiety is released

(A) Phosphate (B) Pyrophosphate
(C) Base (D) Nucleotide.

7. RNA polymerase differs from DNA polymerase in the following character(s)

(A) RNA polymerase does not require a primer
(B) RNA polymerase lacks endo-or exonuclease activity
(C) RNA polymerase does not possess proof-reading activity
(D) All of the above.

8. Termination of transcription is brought about by

(A) Sigma factor (B) RNA polymerase
(C) Rho factor (D) TATA box.

9. In eukaryotic cells, the synthesis of mRNA is carried out by

(A) RNA polymerase I (B) RNA polymerase II
(C) RNA polymerase III
(D) Any one of the above.

10. The primary transcript of mRNA in eukaryotes is heterogeneous nuclear RNA (hnRNA). It is subjected to the following post-transcriptional modifications

(A) 5′-end capping with 7-methylguanosine
(B) Addition of poly A chain
(C) Removal of introns (D) All of the above.

11. The enzyme responsible for reverse transcription

(A) DNA dependent DNA polymerase
(B) DNA dependent RNA polymerase
(C) RNA dependent DNA polymerase
(D) RNA dependent RNA polymerase.

12. Translation refers to the synthesis of

(A) DNA from RNA (B) RNA from DNA
(C) Protein from DNA (D) Protein from RNA.

13. Genetic code is the dictionary of the following nucleotide bases

(A) A, G, C and T (B) A, G, C and U
(C) A, G, T and U (D) A, G, T and T.

14. The initiating codon in protein biosynthesis

(A) UAA (B) UAG (C) UGA (D) AUG.

15. The following are termination or nonsense codons except

(A) AUG (B) UAA (C) UAG (D) UGA.

16. The total number of codons available to code for 20 amino acids

(A) 20 (B) 32 (C) 64 (D) 61.

17. The codon (of mRNA) and anticodon (of tRNA) recognize each other by pairing

(A) 5′→3′ of mRNA with 5′→3′ of tRNA
(B) 3′→5′ of mRNA with 3′→5′ of tRNA
(C) 5′→3′ of mRNA with 3′→5′ of tRNA
(D) In any one of the above manners.

18. Wobble hypothesis is explained by one of the following characteristics of the genetic code

(A) Universality (B) Specificity
(C) Non-overlapping (D) Degenerate.

19. Wobble hypothesis is characterized by recognition of

(A) A single codon by a single tRNA
(B) More than one codon by a single tRNA
(C) A single codon by more than one tRNA
(D) All of the above.

20. The specific information required for the sequence of amino acids in a protein is located on

(A) rRNA (B) mRNA
(C) tRNA (D) hnRNA.

21. The factories for protein synthesis

(A) Mitochondria (B) Nucleus
(C) Ribosomes (D) Cell.

22. During translation, the mRNA is read in 5′→3′ direction and the polypeptide synthesis proceeds from

(A) C-terminal end to C-terminal end
(B) C-terminal end to N-terminal end
(C) N-terminal end to C-terminal end
(d) N-terminal end to N-terminal end.

23. The prokaryote mRNA possesses a specific sequence of nucleotide bases that bind with rRNA of 30S ribosome. This is referred to as

(A) Pribnow box sequence
(B) Hogness box sequence
(C) Shine-Dalgarno sequence
(D) Okazaki fragment.

24. The formation of peptide bonds during protein biosynthesis is catalysed by an enzyme peptidyltranferase. The chemical nature of this enzyme is

(A) Protein (B) DNA
(C) RNA (D) Polysaccharide.

25. One of the following compounds competitively inhibits the enzyme peptidyltransferase and interferes with the elongation of peptide chain

(A) Streptomycin (B) Tetracycline
(C) Erythromycin (D) Chloramphenicol.

26. Post-translational modifications involve the following covalent change(s)

(A) Phosphorylation (B) Hydroxylation
(C) Carboxylation
(D) Any one of the above.

27. The mitochondria possess a limited capability to synthesize a few proteins. These proteins/enzymes are the components of

(A) Citric acid cycle
(B) β-oxidation of fatty acids
(C) Electron transport chain
(D) None of the above.

28. The mitochondrial DNA is inherited from

(A) Father only (B) Mother only

(C) Both father and mother

(D) Either father or mother.

29. The following disorder is a good example of a single base mutation of mitochondrial DNA

(A) Lebers hereditary optic neuropathy

(B) Xeroderma pigmentosum

(C) Fanconi's anemia

(D) Cockayne's syndrome.

30. The replacement of a purine base by another purine base is referred to as

(A) Transition point mutation

(B) Tranversion point mutation

(C) Frame shift mutation

(D) None of the above.

31. Frame shift mutations are characterized by

(A) Insertion or deletion of one or more base pairs in DNA

(B) Insertion of one or more base pairs in DNA

(C) Deletion of one or more base pairs in DNA

(D) Replacement of a purine by a pyrimidine or vice versa.

32. Sickle-cell anemia is a classical example of

(A) Silent mutation (B) Frame shift mutation

(C) Nonsense mutation

(D) Missense mutation.

33. One of the following mutations causes the synthesis of proteins with several altered amino acids and/or prematurely terminated proteins.

(A) Silent mutation

(B) Frame shift mutation

(C) Nonsense mutation

(D) Missense mutation.

34. The following disease is a good example of mutations in mitocondrial DNA

(A) Pearson syndrome

(B) Leber's hereditary optic neuropathy

(C) Hunter syndrome (D) Gaucher's disease

Answers

1. **(B)** 2. **(D)** 3. **(C)** 4. **(A)** 4. **(B)** 5. **(B)** 6. **(B)** 7. **(D)** 8. **(C)** 9. **(B)** 10. **(D)** 11. **(C)** 12. **(D)** 13. **(B)** 14. **(D)** 15. **(A)** 16. **(D)** 17. **(C)** 18. **(D)** 19. **(B)** 20. **(B)** 21. **(C)** 22. **(C)** 23. **(C)** 24. **(C)** 25. **(D)** 26. **(D)** 27. **(C)** 28. **(B)** 29. **(A)** 30. **(A)** 31. **(A)** 32. **(D)** 33. **(B)** 34. **(B)**

For related clinical case studies, refer 1.1, 7.3, 10.2 and 10.6

VIVA/Short Questions

1. What do you understand by transcription?

A process in which RNA is synthesized from DNA. It occurs in 3 stages - initiation, elongation and termination.

2. What is the role of RNA polymerase?

RNA polymerase is responsible for the synthesis of RNA from DNA.

3. What is a Pribnow (TATA) box?

It is a sequence (TATAAT) on the non-coding DNA strand. The sigma factor of RNA polymerase recognizes Pribnow box for the initiation of transcription.

4. How does RNA polymerase differ from DNA polymerase?

RNA polymerase does not require a primer and does not possess proof-reading activity while DNA polymerase has both these characteristics.

5. What are palindromes?

A palindrome is a word that reads alike forward and backward. e.g., madam, rotor.

6. What is the importance of palindromes in transcription?

The presence of palindromes in the base sequence of DNA template (same when read in opposite direction) in the termination region is known. As a result of this, the newly synthesized RNA folds to form hairpins (due to complementary base pairing) that cause termination of transcription.

7. What do you know about the RNA polymerases of eukaryotes?

The nuclei of eukaryotes cells possess 3 distinct RNA polymerases. RNA polymerase I being responsible for the synthesis of ribosomal RNAs, RNA polymerase II synthesizes mRNAs, while RNA polymerase III participates in the formation of tRNAs.

8. What are the promoter sites for transcription in eukaryotes?

Hogness box (or TATA box) and CAAT box.

9. What is heterogenous nuclear RNA?

The primary mRNA transcript produced in eukaryotes is referred to as heterogenous nuclear RNA (hnRNA).

10. What do you know about post- transcriptional modifications?

The RNAs produced during transcription undergo many alterations - terminal base additions, base modifications, splicing etc. This process is required to convert RNAs into the active forms.

11. What are introns and what are exons?

Introns are the intervening nucleotide sequences in mRNA which do not code for proteins. On the other hand, exons of mRNA possess genetic code and are responsible for protein synthesis.

12. Name some inhibitors of transcription.

Actinomycin D, rifampin and α-amanitin.

13. What do you know about reverse transcription?

The process of synthesis of DNA from RNA (catalysed by reverse transcriptase) is referred to as reverse transcription. Some of the viruses known as retroviruses, possess RNA as the genetic material. In these viruses, DNA is synthesized from RNA.

14. What is translation?

Biosynthesis of a protein or a polypeptide in a living cell.

15. What determines the sequence of amino acids in a protein?

The nucleotide base sequence of mRNA (i.e., genetic code).

16. What is genetic code?

The three nucleotide (triplet) base sequences in mRNA that act as code words for amino acids in protein constitutes genetic code or simply codons.

17. What is the composition of codons?

Codons are composed of four nucleotide bases, namely the purines-adenine (A) and guanine (G) and the pyrimidines-cytosine (C) and uracil (U). These four bases produce 64 combinations of three base codons to code for 20 amino acids found in proteins.

18. What do you understand by nonsense codons?

Nonsense codons or termination codons act as stop signals in protein synthesis. These are UAA, UAG and UGA.

19. What are initiating codons?

The codon AUG (sometimes GUG) initiate the polypeptide synthesis.

20. What are the characteristics of genetic code?

The genetic code is universal, specific, non-overlapping and degenerate.

21. What do you understand by degeneracy of a codon?

The occurrence of multiple codons for amino acids is referred to as degeneracy. This is obvious from the fact that there are 61 codons for only 20 amino acids.

22. What is wobble hypothesis?

Wobble hypothesis is the phenomenon in which a single tRNA can recognize more than one codon. This explains the existence of multiple codons for a single amino acid i.e., degeneracy of genetic code.

23. Name the important raw materials required for protein biosynthesis.

Amino acids, ribosomes, mRNA, tRNAs, energy sources (ATP, GTP), protein factors etc.

24. What is the role of ribosomes in translation?

Ribosomes are the centres or factories for protein synthesis (translation).

25. What is a polyribosome (polysome)?

The term polyribosome is used when several ribosomes simultaneously translate on a single mRNA.

26. What is the role of tRNAs in protein synthesis?

tRNAs carry the amino acids and hand them over to the growing peptide chain. Each tRNA has a three nucleotide base sequence-the anticodon which is responsible to recognize the codon of mRNA for protein synthesis.

27. How are the amino acids activated and made ready for protein synthesis?

Each one of the amino acids is first attached to the enzyme aminoacyl tRNA synthetase which is then transferred to tRNA to form aminoacyl tRNA.

28. What do you understand by polycistronic mRNA and monocistronic mRNA?

The prokaryotic mRNA is polycistronic since a single mRNA has many coding regions that code for different polypeptides. In contrast, eukaryotic mRNA is monocistronic, since it codes for a single polypeptide.

29. What is Shine-Dalgarno sequence?

The prokaryote mRNA possesses a specific sequence of nucleotide bases near 5′-end (5′-UAAGGAGG-3′) which is known as Shine-Dalgarno sequence. This is required for the proper selection of initiation codon.

30. What is an A site and what is a P site in a ribosome?

A site of the ribosome corresponds to the place where an incoming amino acid (as aminoacyl tRNA) is deposited. P site refers to the place where the growing peptide chain is transferred.

31. How is the peptide bond formed in protein synthesis?

The enzyme peptidyltransferase catalyses the formation of peptide bond. The activity of the enzyme lies on the rRNA (not protein) of 50s ribosome. This rRNA is referred to as ribozyme.

32. Name some inhibitors of protein synthesis?

Streptomycin, tetracycline, puromycin, chloramphenicol, erythromycin, diphtheria toxin.

33. How does puromycin inhibit protein biosynthesis?

The antibiotic puromycin has structural resemblance to aminoacyl tRNA. It enters the A site and gets incorporated into the growing peptide chain and causes its release. Puromycin can inhibit protein synthesis in prokaryotes and eukaryotes.

34. What are post-translational modifications of proteins?

The changes that take place in the polypeptides after the initiation of their synthesis or most frequently after the protein synthesis in completed. These include proteolytic degradation and covalent changes. Phosphorylation, hydroxylation and glycosylation are the important covalent modifications.

35. What do you know about the role of vitamins in post-translational modifications of proteins?

Vitamin C is required for the hydroxylation of proline and lysine residues during collagen formation. Vitamin K is necessary for the carboxylation of glutamic acid residues in certain clotting factors (II, VII, IX, X).

36. What are the characteristics of mitochondrial DNA?

Mitochondrial DNA (mtDNA) is circular in and contains about 16,000 nucleotide bases. It is inherited from the mother. mtDNA is subjected to high rate of mutations (10 times more than nuclear DNA).

37. Define mutation.

Mutation refers to a change in the DNA structure of a gene.

26. Regulation of Gene Expression

MCQs

1. The term genome refers to the

(A) Functional unit of gene
(B) Total genetic information contained in a cell
(C) Structural genes only
(D) Operator genes only.

2. The human genome contains about

(A) 1,000 genes (B) 10,000 genes
(C) 20,000 genes (D) 1,000,000 genes.

3. The gene responsible for insulin synthesis is present in

(A) Pancreas
(B) β-cells of pancreas only
(C) Pancreas and intestine
(D) All the cells of the body.

4. One of the following enzymes is synthesized by an inducible gene

(A) Pyruvate dehydrogenase
(B) Fumarase (C) Tryptophan pyrrolase
(D) Pyruvate kinase.

5. One cistron refers to the fragment of the DNA that can synthesize

(A) A protein (B) A polypeptide
(C) An isoenzyme (D) An enzyme.

6. The structural genes Z, Y and A in lac operon respectively synthesize

(A) β-galactosidase, galactose permease and galactoside acetylase
(B) Galactose permease, galactoside acetylase and β-galactosidase
(C) Galactoside acetylase, galactose permease and β-galactosidase
(D) β-galactosidase, galactoside acetylase and galactose permease.

7. The enzyme RNA polymerase binds to

(A) Regulatory gene (B) Promoter site
(C) Operator gene (D) Structural gene A.

8. Lac repressor protein is synthesized by

(A) Regulatory gene (B) Operator gene
(C) Structural gene (D) Promoter site.

9. In the lac operon, the repressor molecule acts as a

(A) Promoter of gene expression
(B) Positive regulator (C) Negative regulator
(D) None of the above.

10. In the lac operon, lactose acts by

(A) Activating the repressor
(B) Inactivating the repressor
(C) Promoting the operator gene
(D) Inhibiting the expression of structural genes.

11. One of the following compounds interferes with the induction of lac operon

(A) Lactose (B) Galactose
(C) Glucose (D) Fructose.

12. The binding of RNA polymerase to promoter site requires the presence of a catabolite gene activator protein (CAP) bound to

(A) Cyclic AMP (B) Cyclic GMP
(C) ATP (D) Glucose.

13. The malignant cells develop resistance to methotrexate (drug resistance) by the following action in relation to the synthesis of dihydrofolate reductase

(A) Gene rearrangement
(B) Degradation of mRNA
(C) Gene amplification
(D) Processing of mRNA.

14. The body possesses an enormous capacity to synthesize a wide range (about 10 billion) of antibodies. This is attributed to

(A) Gene amplification
(B) Gene rearrangement
(C) Amplification of mRNA
(D) All of the above.

15. The control of transcription is brought about by the following motif(s)

(A) Helix-turn-helix (B) Zinc finger
(C) Leucine zipper (D) All of the above.

Answers

1. **(B)** 2. **(C)** 3. **(D)** 4. **(C)** 5. **(B)** 6. **(A)** 7. **(B)** 8. **(A)** 9. **(C)** 10. **(B)** 11. **(C)** 12. **(A)** 13. **(C)** 14. **(B)** 15. **(D)**.

VIVA/Short Questions

1. What is genome?

The term genome refers to the total genetic information contained in a cell.

2. What do you know about human genome?

The human genome is complex with 23 pairs of (diploid) chromosomes containing 6 billion (6×10^9) base pairs in DNA, with an estimated 20,000 genes.

3. What is importance of gene regulation?

The living cells possess a remarkable property to adapt to changes in the environmet by regulating the gene expression. For instance, insulin is synthesized by specialized cells of pancreas and not by the cells of other organs (say kidney, liver), although the nuclei of all the cells of the body contain the insulin genes. The regulation of the gene expression is absolutely essential for growth, development, differentiation and the very existence of an organism.

4. What are the types of gene regulations?

There are two types of gene regulations — positive and negative. In the positive gene regulation, the gene expression is increased by regulatory element, while the regulatory element decreases the gene regulation in negative regulation.

5. What are constitutive genes?

Constitutive genes (or house-keeping genes) are expressed at more or less constant rate in almost all the cells and further, they are not subjected to regulation. The products (proteins) of constitutive genes are required all the time in a cell. e.g., enzymes of citric acid cycle.

6. What do you understand by inducible genes?

The concentration of the proteins synthesized by inducible genes is regulated by various molecular signals. An inducer increases the expression of these genes while a repressor decreases e.g., tryptophan induces the synthesis of tryptophan pyrrolase in the liver.

7. What is a cistron?

A cistron is the smallest unit of genetic expression. It is the fragment of DNA coding for the subunit of a protein molecule.

8. What is one cistron — one subunit concept?

One cistron (the functional fragment of DNA) can code for a subunit (polypeptide) of a protein. And this is one cistron — one subunit concept.

9. What is operon?

Operon is the coordinated unit of genetic expression in bacteria.

10. Who introduced the concept of operon?

Jacob and Monod in 1961 (Nobel Prize 1965).

11. What is lac operon?

Lac operon is the coordinated expression of the genes in relation to lactose metabolism in *E. coli*.

12. How is lac operon composed?

The lac operon consists of a regulatory gene (I), operator gene (O) and three structural genes (Z, Y, A), besides the promoter site (P). The structural genes Z, Y and A respectively code for the enzymes β-galactosidase, galactose permease and galactoside acetylase.

13. What is polycistronic mRNA?

An mRNA coding for more than one protein (or polypeptide) is known as polycistronic mRNA. Prokaryotic organisms contain a large number of polycistronic mRNAs e.g., polycistronic mRNA of lac operon (for genes Z, Y, A).

14. What are gratuitous inducers?

The structural analogs of lactose which can induce the lac operon but are not the substrates for β-galactosidase are the gratuitous inducers e.g., isopropylthiogalactoside.

15. What are the major mechanisms of eukaryotic gene regulation?

Eukaryotic gene expression and its regulation are highly complex and the knowledge in this regard is fragmentary. Some of the important methods are gene amplification, gene rearrangement, synthesis of primary RNA transcript, processing of RNA, transport of mRNA from nucleus to cytoplasm and degradation of mRNA.

16. How does gene amplification occur?

The expression of a gene is increased several fold. This is commonly observed during the developmental stages of eukaryotic organisms. In humans, the development of drug resistance to methotrexate in the malignant cells occurs by amplifying the genes coding for dihydrofolate reductase.

17. What is the molecular basis for the ability of the body to produce wide range of antibodies?

It is estimated that the human body can produce about 10 billion antibodies in response to antigen stimulation. It is explained on the basis of gene rearrangement.

18. What are enhancers and silencers?

Certain DNA segments promote the synthesis of RNA which are known as enhancers. On the other hand, some genes referred to as silencers, diminish the transcription process.

19. Name the regulatory proteins that bind to specific DNA segments and control transcription?

The specific interaction of proteins to DNA is brought about by three unique motifs—helix-turn-helix, zinc finger and leucine zipper.

27. Recombinant DNA and Biotechnology

MCQs

1. The group of enzymes used for splicing of specific fragments of DNA from the genome

(A) Restriction exonucleaes
(B) Restriction endonucleases
(C) Restriction transferases
(D) Hydrolases.

2. The number of base pairs each restriction endonuclease can recognise

(A) 1-4 (B) 4-7 (C) 7-11 (D) 11-14.

3. The vector(s) employed in recombinant DNA technology

(A) Plasmids (B) Bacteriophages
(C) Cosmids (D) All of them.

4. Plasmids contain a

(A) Circular single stranded DNA
(B) Linear single stranded DNA
(C) Circular double stranded DNA
(D) Linear double stranded DNA.

5. Plasmids in bacteria provide resistance against

(A) Viruses (B) Antigens
(C) Antibiotics (D) All of them.

6. Plasmids, the most commonly used vectors can accept DNA fragments of

(A) 2-6 kilobase length
(B) 6-10 kilobase length
(C) 10-20 kilobase length
(D) 20-50 kilobase length.

7. The enzyme employed for joining the DNA fragments to finally produce recombinant DNA

(A) DNA polymerase (B) DNA ligase
(C) Endonuclease (D) Exonuclease.

8. Complementary DNA (cDNA) contains copies of DNA sequence that are present in the

(A) DNA (B) hnRNA
(C) mRNA (D) tRNA.

9. The technique employed for the amplification of a selected DNA fragment

(A) Replication (B) Recombination
(C) Polymerase chain reaction
(D) Transformation.

10. The blotting techniques respectively used for the identification of DNA, RNA and protein

(A) Northern blot, Southern blot and Western blot
(B) Southern blot, Northern blot and Western blot
(C) Western blot, Northern blot and Sourthern blot
(D) Southern blot, Western blot and Northern blot.

11. Restriction fragment length polymorphisms (RFLPs) refer to the slight variations in the base sequence of

(A) DNA (B) mRNA (C) tRNA (D) rRNA.

12. The application(s) of recombinant DNA technology—

(A) Production of proteins in abundance
(B) Molecular analysis of diseases
(C) Gene therapy (D) All of the above.

13. **The first successful gene therapy in humans was carried out for the enzyme**
    (A) Lipoprotein lipase
    (B) Glucose 6-phosphatase
    (C) Adenosine deaminase
    (D) Tyrosinase.

Answers

1. **(B)** 2. **(B)** 3. **(D)** 4. **(C)** 5. **(C)** 6. **(B)** 7. **(B)** 8. **(C)** 9. **(C)** 10. **(B)** 11. **(A)** 12. **(D)** 13. **(C)**.

VIVA/Short Questions

1. **What is genetic engineering?**

The manipulation of genetic material (DNA) to achieve the desired goal.

2. **What is a recombinant DNA or chimeric DNA?**

Insertion of the selected DNA into a cloning vector (e.g. a plasmid) to create a recombinant DNA or chimeric DNA.

3. **What do you understand by restriction endonuclease?**

The bacterial enzymes that can cut/split DNA (from any source) at specific sites.

4. **Name the most commonly used gene transfer methods.**

Transformation, conjugation, electroporation and lipofection, and direct transfer of DNA.

5. **What are blotting techniques?**

The laboratory techniques used for identification of desired DNA or RNA fragments or proteins.

6. **Name the blotting techniques and the corresponding nucleic acids identified.**

Southern blotting (for DNA), Northern blotting (for RNA), Dot blotting (DNA/RNA).

7. **What is the purpose of western blotting technique?**

Identification of proteins.

8. **What are DNA chips (microarrays)?**

Recent developments for DNA sequencing as a result of advances made in automation and miniaturization.

9. **What is polymerase chain reaction (PCR)?**

A laboratory (in vitro) technique for generating large quantities of a specified DNA. PCR is a cell-free amplification technique for synthesizing multiple identical copies (billions) of any DNA of interest.

10. **Which PCR is used for the diagnosis of covid-19?**

Real-time PCR (RT-PCR).

11. **What is site-directed mutagenesis?**

The technique for generating amino acid coding changes in the DNA (gene).

12. **What is genetic detective?**

DNA fingerprinting is the present day genetic detective in the practice of modern medical forensics.

13. **What do you understand by restriction fragment length polymorphisms (RFLPs)?**

DNAs that serves as a markers for mapping a specified gene. RFLPs are located randomly throughout a person's chromosomes and have no apparent function.

14. **What are single nucleotide polymorphisms (SNPs)?**

SNPs represent the positions in the genome where some individuals have one nucleotide (e.g. G) while others have a different nucleotide (e.g. C).

15. **What are recombinant veccines?**

New generation vaccines produced by recombinant DNA technology.

16. **Name the first recombinant vaccine produced.**

Hepatitis B vaccine in 1987.

17. **What are the major vaccines developed by different countries for covid-19.**

Covaxin (inactivated virus), developed by India.

Covishield/Astrazenca U.K, Sweden and India.

Moderna (mRNA based) developed by US.

Sputnik V (viral vector-modified adeno), developed by Russia.

Pfizer and Biotech (mRNA based) developed by US and Germany.

18. **What is transgenesis?**

The phenomenon of introduction of exogeneous DNA into the genome to create and maintain a stable heritable character.

19. **Which is the animal model developed for cancer?**

Oncomouse

20. **What is ELSI of biotechnology?**

ELSI is the short form to represent the ethical, legal and social implications of biotechnology.

CURRENT TOPICS VI

28. Human Genome Project

MCQs

1. **What proportion of genome codes for proteins?**
   (A) 5-6% (B) 1.1-1.5%
   (C) 3.5-4.5% (D) 15-16%.
2. **What percentage of human DNA is identical to that of chimpanzee?**
   (A) 50% (B) 75% (C) 90% (D) 98%.
3. **The member of genes estimated to be present humans**
   (A) 10,000-15,000 (B) 20,000-25,000
   (C) 30,000-35,000 (D) 40,000-45,000.
4. **One of the following country was not a participant in the international venture of in human genone project.**
   (A) China (B) Japan
   (C) Australia (D) France.

Answers

1. **(B)** 2.**(D)** 3. **(B)** 4.**(C)**.

VIVA/Short Questions

1. **What do you understand by human genome project (HGP)?**

Determination of the nucleotide sequence of the entire human nuclear genome.

2. **When was the draft sequence of human genome announced.**

26th June 2000; published in February 2001 in scientific journal, Nature.

3. **What are the different maps constructed for each chromosome in human genome project?**

Cytogenetic map, gene linkage map, restriction fragment map

4. **What is single nucleotide polymorphism (SNP)?**

SNPs represent positions in the genome where some individuals have one nucleotide (i.e. an A), and others have a different nucleotide (i.e. a G).

5. **What is the significance of SNPs?**

The differences between individuals are largely due to single nucleotide polymorphisms (SNPs).

6. **What do you understand by junk DNA?**

A huge portion of the genome is composed of introns, and intergenic sequences (junk DNA).

29. Gene Therapy

MCQs

1. **The following are employed as vectors in gene therapy except**
   (A) Bacteria (B) Viruses
   (C) Bone marrow cell
   (D) Human artificial chromosome.
2. **Severe combined immunodeficiency (SCID) is mostly associated with**
   (A) Hepatic cells (B) Renal cells
   (C) B-lymphocytes (D) T-Lymphocytes
3. **The following enzyme, produced by a suicide gene, is used for the treatment of cancers**
   (A) Adenosine deaminase
   (B) Thymidine kinase (C) DNA helicase
   (D) Telomerase.
4. **Antisense gene therapy refers to**
   (A) Inhibition of translation
   (B) Blocking of transcription
   (C) Inhibition of replication
   (D) Enhanced transcription.
5. **Tumor necrosis factor (TNF) gene is for the treatment of**
   (A) Prostate cancer (B) Ovarian cancer
   (C) Malignant melanoma
   (D) Breast cancer

Answers

1. **(A)** 2.**(D)** 3. **(B)** 4.**(A)** 5. **(C)**

VIVA/Short Questions

1. **What is gene therapy?**

The process of inserting genes into cells to treat diseases.

2. **What to you understand by gene augmentation therapy?**

A DNA is inserted into the genome to replace the missing gene product.

3. **What is gene inhibition therapy?**

The antisense gene inhibits the expression of the dominant gene.

4. What is done in somatic cell gene therapy?

Insertion of a fully functional and expressible gene into a target somatic cell to correct a genetic disease permanently.

5. What do you understand by *ex vivo* gene therapy?

Transfer of genes in cultured cells (e.g., bone marrow cells).

6. What is *in vivo* gene therapy?

Direct delivery of genes into the cells.

7. Name some vectors used in gene therapy.

Viruses, human artificial chromosome, bone marrow cells.

8. What is severe combined immunodeficiency (SCID)?

Patients of SCID lack the enzyme adenosine deaminase and suffer from infectious diseases and die at an young age.

9. Name the first disease for which gene therapy was carried out in humans.

SCID to correct the deficiency of the enzyme, adenosine deaminase (ADA).

10. What is antigene therapy?

Blocking transcription using a single-stranded nucleotide sequence (antigene oligonucleotide).

11. What is antisense therapy?

Inhibition of translation by using a single-stranded nucleotide (antisense oligonucleotide).

12. What do you understand by nucleic acid therapy?

Use of DNA or RNA molecules for therapeutic purposes.

13. What are peptide nucleic acids (PNAs)?

Artificial analogs of nucleic acid with a polypeptide backbone.

31. Metabolism of Xenobiotics

MCQs

1. The site of the metabolism of foreign compounds

(A) Brain (B) Liver
(C) Heart (D) Kidney.

2. Phase I reactions of detoxification are the following reactions except

(A) Oxidation (B) Reduction
(C) Hydrolysis (D) Conjugation.

3. Phase II reactions of detoxification involve

(A) Oxidation (B) Reduction
(C) Hydrolysis (D) Conjugation.

4. Conversion of benzaldehyde to benzoic acid is an example of

(A) Oxidation (B) Reduction
(C) Hydrolysis (D) Conjugation.

5. The oxidation reactions of detoxification are catalysed by

(A) Cytochrome C
(B) Cytochrome oxidase
(C) Cytochrome P_{450} (D) Cytochrome a.

6. Cytochrome P_{450} is dependent on the coenzyme

(A) NADH (B) NADPH
(C) $FADH_2$ (D) $FMNH_2$.

7. The following statement about cytochrome P_{450} is not true

(A) It is hemoprotein found in microsomes
(B) It exists in multiple species
(C) It is an inducible enzyme
(D) It participates in the conjugation reaction.

8. Conversion of picric acid to picramic acid in detoxification is an example of

(A) Oxidation (B) Reduction
(C) Hydrolysis (D) Conjugation.

9. The detoxification of aspirin involves

(A) Oxidation (B) Reduction
(C) Hydrolysis (D) All of the above.

10. The most common form of conjugation under detoxification involves the following agent

(A) Glucuronic acid (B) Glycine
(C) Cysteine (D) Glutamine.

11. The amino acid involved in the formation of hippuric acid

(A) Glutamic acid (B) Glycine
(C) Cysteine (D) Glutamine.

12. Cysteine, utilized as conjugating agent in detoxification, is derived from

(A) Methionine (B) Cystine
(C) Glutathione (D) Carnitine.

13. The highly toxic cyanides are conjugated with the following compound to form less toxic compounds

(A) Active sulfate (B) Thiosulfate
(C) Active methionine (D) Cysteine.

14. The drug BAL (British antilewisite) developed during the world war II can relieve the toxic symptoms of the following metal(s)

(A) Arsenic (B) Mercury
(C) Cadmium (D) All of the above.

Answers

1. **(B)** 2. **(D)** 3. **(D)** 4. **(A)** 5. **(C)** 6. **(B)** 7. **(D)** 8. **(B)** 9. **(C)** 10. **(A)** 11. **(B)** 12. **(C)** 13. **(B)** 14. **(D)**.

For related clinical case study, refer 20.1

VIVA/Short Questions

1. What do you understand by the term detoxification?

The series of biochemical reactions occurring in the body to convert the foreign (often toxic) compounds to non-toxic or less toxic and more easily excretable forms.

2. What are the terms recently in use to replace detoxification?

Biotransformation or metabolism of xenobiotics or metabolisim of foreign compounds.

3. Where does detoxification occur?

Mainly in the liver.

4. What is the mechanism of detoxification?

Phase I reactions involving oxidation, reduction and hydrolysis. In ***phase II***, conjugating reactions occur involving compounds such as glucuronic acid, amino acids (glycine), glutathione.

5. Give some examples of oxidation reactions of detoxification.

Conversion of methanol to formic acid; ethanol to acetic acid; benzaldehyde to benzoic acid, toluene to benzoic acid.

6. What do you know about cytochrome P_{450}?

The oxidation reactions of detoxification are largely catalysed by mixed function oxidase or cytochrome P_{450} which is found in association with microsomes.

7. Give some examples of hydrolysis reactions of detoxification.

Aspirin to salicylic acid and acetic acid; atropine to tropic acid and tropine.

8. What do you understand by conjugation in relation to detoxification?

Conjugation is the process in which a foreign compound combines with a substance produced in the body. Conjugation may occur either directly or after phase I reactions of detoxification.

9. Name the different conjugating agents in the metabolism of xenobiotics.

Glucuronic acid, glycine, cysteine, glutamine, methyl group, sulfate, acetic acid and thiosulfate.

10. Give some examples of glycine conjugation.

Many aromatic corboxylic acids (e.g., benzoic acid, phenylacetic acid) are conjugated with glycine. Hippuric acid is formed when glycine is conjugated with benzoic acid.

11. How is cyanide detoxified?

Highly toxic cyanides get conjugated with thiosulfate to form less toxic thiocynates.

12. What do you know about the detoxification by drugs?

The toxic effects of certain metals such as arsenic, mercury and cadmium could be overcome by administering British antilewisite (BAL). BAL was developed during the world war II and was used as a detoxifying agent for certain war poisons.

32. Prostaglandins and Related Compounds

MCQs

1. The term eicosanoids is used to represent

(A) Prostaglandins
(D) Prostacyclins
(C) Thromboxanes and leukotrienes
(D) All of the above.

2. Prostaglandins are not synthesized in

(A) Prostate gland (B) Liver
(C) Kidney (D) Erythrocytes.

3. The parent compound to which all the prostaglandins are structurally related

(A) Arachidonic acid
(B) Prostanoic acid
(C) Eicosapentaenoic acid
(D) Oleic acid.

4. The ring structure present in the prostaglandins and related compounds

(A) Benzene (B) Pyridine
(C) Cyclopentane (D) Indole.

5. The precursor for the biosynthesis of prostaglandins

(A) Linoleic acid (B) Linolenic acid
(C) Arachidonic acid (D) Prostanoic acid.

6. **The fatty acid precursor for prostaglandin synthesis is released from the membrane bound**
   (A) Triacylglycerols (B) Cholesterol esters
   (C) Glycolipids (D) Phospholipids.
7. **The compound that irreversibly inhibits the cyclooxygenase of prostaglandin synthesis**
   (A) Cortisol (B) Phenylbutazone
   (C) Aspirin (D) Indomethacin.
8. **Cortisol inhibits prostaglandin synthesis by inhibiting the enzyme**
   (A) Phospholipase A_1 (B) Phospholipase A_2
   (C) Phospholipase C (D) Cyclooxygenase.
9. **Prostaglandins (PGs) differ from the hormones in the following character(s)**
   (A) PGs are produced in almost all tissues
   (B) PGs are not stored
   (C) PGs act as local hormones
   (D) All of the above.
10. **Prostaglandins perform the following function(s)**
    (A) Regulation of blood pressure
    (B) Inflammatory reactions
    (C) Regulation of gastric secretion
    (D) All of the above.
11. **Platelet aggregation is inhibited by**
    (A) Prostaglandin E_2 (B) Prostacyclins
    (C) Thromboxanes (D) Leukotrienes.
12. **The following compound is an intermediate in the synthesis of leukotrienes**
    (A) Arachidonic acid (B) Prostanoic acid
    (C) Eicospentaenoic acid
    (D) 5-hydroperoxy eicosatetraenoic acid.
13. **The compounds that promote platelet aggregation and thrombosis**
    (B) Prostacyclins (B) Leukotrienes
    (C) Thromboxanes (D) Prostaglandin A.
14. **The marine lipids are rich in the following fatty acid, which offers protection against heart attacks**
    (A) Linoleic acid (B) Linolenic acid
    (C) Eicosatetraenoic acid
    (D) Eicosapentaenoic acid.

Answers

1. **(D)** 2. **(D)** 3. **(B)** 4. **(C)** 5. **(C)** 6. **(D)** 7. **(C)** 8. **(B)** 9. **(A)** 10. **(D)** 11. **(B)** 12. **(D)** 13. **(C)** 14. **(D)**.

VIVA/Short Questions

1. **What are eicosanoids?**

20 carbon containing fatty acids. Prostaglandins and their related compounds — prostacyclins, thromboxanes and leukotrienes are collectively known as eicosanoids.

2. **What are prostaglandins?**

Prostaglandins (PGs) are derivatives of a 20 — carbon fatty acid namely prostanoic acid.

3. **How are prostaglandins synthesized?**

Arachidonic acid, derived from the membrane phospholipids, is the precursor.

4. **Name the inhibitors of prostaglandin synthesis.**

Corticosteroids (e.g., cortisol), aspirin, indomethacin, phenylbutazone and ibuprofen.

5. **What are the biochemical actions of prostaglandins?**

Act as local hormones in their function. PGs decrease blood pressure, cause inflammation, inhibit gastic secretion, decrease immunological function.

6. **What is the role of prostaglandins in relation to platelet aggregation and thrombosis?**

Prostacyclins (PGI_2) inhibit platelet aggregation. On the other hand, thromboxanes (TXA_2) and prostaglanadin E_2 promotes platelet aggregation and blood clotting that might lead to thrombosis.

7. **What are the therapeutic applications of prostaglandins?**

Used in the treatment of gastric ulcers, hypertension, thrombosis, asthma etc. PGs are also employed in the medical termination of pregnancy, prevention of conception, induction of labour etc.

8. **What do you know about leukotrienes?**

Synthesized by the lipooxygenase pathway of arachidonic acid. Leukotrienes have been implicated in asthma, inflammatory reactions, hypersensitivity (allergy) and heart attacks.

9. **What is the importance of fish oils in relation to heart diseases?**

Fish oils with UFA prevent platelet aggregation and thrombosis.

33. Biological Membranes and Transport

MCQs

1. **The biological membranes are composed of the following compounds except**
   (A) Lipids (B) Proteins
   (C) Nucleic acids (D) Carbohydrates.

2. According to the fluid mosaic model, the biological membranes are composed of

(A) Lipid monolayer (B) Lipid bilayer
(C) Protein-lipid layer (D) Protein bilayer.

3. The transport across the membrane which depends only on the cancentration gradient

(A) Passive diffusion
(B) Facilitated diffusion (C) Active transport
(D) None of the above.

4. The diffusion of solutes along the concentration gradient through the mediation of specific carrier protein represents

(A) Passive diffusion (B) Facilitated diffusion
(C) Active transport
(D) Any one of the above.

5. Active transport is characterized by

(A) Movement of solute against concentration gradient
(B) Supply of ATP as energy source
(C) Involvement of protein carriers
(D) All of the above.

6. Na^+-K^+ pump is responsible for the maintenance of

(A) Equal concentration of Na^+ and K^+ inside cells
(B) High intracellular K^+ and low Na^+ concentration
(C) Low intracellular K^+ and high Na^+ concentration
(D) High intracellular K^+ and low Cl^- concentration.

7. In the operation of Na^+-K^+ pump, for every 3 Na^+ ions pumped from inside to outside of the cell, number of K^+ ions that enter the cell

(A) 1 (B) 2 (C) 3 (D) 4.

8. The Na^+-K^+ pump is inhibited by

(A) Fluoride (B) Iodoacetate
(C) Ouabain (D) Fluoroacetate.

9. The simultaneous transport of Na^+ and glucose to the intestinal mucosal cells from the gastrointestinal tract is an example for the following type of transport

(A) Uniport system (B) Symport system
(C) Antiport system (D) None of the above.

10. A good example for antiport system of transport

(A) Transport of glucose to RBC
(B) Transport of Na^+ and glucose to intestinal mucosal cells from the gut
(C) Exchange of Cl^- and HCO_3^- in RBC
(D) None of the above.

11. The protein present in the membrane coated pits that participates in endocytosis (intake of macromolecules by the cells)

(A) Na^+-K^+ ATPase (B) Clathrin
(C) Ouabain (D) Cytochrome P_{450}.

12. The cellular organelles actively involved in exocytosis (release of macromolecules from the cells)

(A) Mitochondria (B) Microsomes
(C) Golgi apparatus (D) Nuclei.

Answers

1. **(C)** 2. **(B)** 3. **(A)** 4. **(B)** 5. **(D)** 6. **(B)** 7. **(B)** 8. **(C)** 9. **(B)** 10. **(C)** 11. **(B)** 12. **(C)**.

VIVA/Short Questions

1. What are the major chemical constituents of biological membranes?

Lipids, proteins and carbohydrates. Among the lipids, phospholipids, glycolipids and cholesterol are found in animal membranes.

2. What are the different mechanisms of transport across the membranes?

Passive diffusion, facilitated diffusion and active transport. Passive diffusion is simple and depends on the concentration gradient while facilitated diffusion occurs through the mediation of carrier or transport proteins. Active transport on the other hand, occurs against a concentration gradient which is dependent on the supply of energy (ATP) e.g., Na^+—K^+ pump.

3. What is the importance of Na^+—K^+ pump?

The Na^+—K^+ pump is responsible for the maintenance of high K^+ and low Na^+ concentrations in the cells, a prerequisite for the survival of the cells.

4. What are the transport systems operative through the membrane?

Three types of transport systems occur for transport of solutes through the membrane; (i) ***uniport*** systems involves the movement of a single molecule through the membrane e.g., transport of glucose to the RBC; (ii) ***symport*** systems — the simultaneous transport of different molecules in the same direction. e.g., transport of glucose and Na^+ to the intestinal mucosal cells from the gut; (iii) ***antiport*** system involving the simultaneous transport of two different molecules in the opposite direction e.g., Cl^- and HCO_3^- in the RBC.

5. What do you understand by cotransport system?

The symport and antiport systems are collectively referred to as contransport system.

6. How does transport of macromolecules occur through the membrane?

Endocytosis (intake of macromolecules) and exocytosis (release of macromolecules) are two independent processes operative for the transport of macromolecules (proteins, polynucleotides) through the membranes.

34. Free Radicals and Antioxidants

MCQs

1. One of the following is not a free radical
(A) Nitric oxide (B) Oxygen
(C) Superoxide (D) Hydrogen peroxide

2. Malondialdehyde (MDA) is used as a biochemical marker for
(A) Fragmentation of proteins
(B) Breaking of DNA strands
(C) Glycation of proteins
(D) Lipid peroxidation.

3. The cellular organelle that is closely associated with the free radical scavenging
(A) Nucleus (B) Golgi appearatus
(C) Peroxisomes (D) Lysosomes.

4. Respiratory burst occurs in the following cells
(A) Erythrocytes (B) Neutrophils
(C) Basophils (D) Macrophages.

5. One of the following is not a metabolic antioxidant
(A) Glutathione (B) Vitamin C
(C) Uric acid (D) Bilirubin.

Answers

1. **(B)** 2.**(D)** 3. **(C)** 4.**(D)** 5. **(B)**.

For related clinical case studies, refer 10.5 and 13.3

VIVA/Short Questions

1. What is a free radical?

A molecule or a molecular species that contains one or more unpaired electrons, and is capable of independent existence.

2. Give some examples of free radicals.

O_2^- (Superoxide), H_2O_2 (hydrogen peroxide) OH^- (hydroxyl radical), nitric oxide (NO^-).

3. What are the sources for generation of free radicals?

Normal biological processes, environmental effects.

4. Name some diseases where free radicals have been implicated.

Cardiovascular diseases (CHD), cancer, inflammatory diseases, respiratory diseases.

5. What are the antioxidant enzymes that can scavange the effects of free radicals.

Superoxide dismutase, catalase, glutathione peroxidase, glutathione reductase.

6. Name some nutrient antioxidants.

Tocopherols (vitamin E), ascorbic acid (vitamin C), carotenoids, β-carotene, selenium, α-lipoic acid, caffeine.

35. Environmental Biochemistry

MCQs

1. The most dangerous pollutant gas to man
(A) Sulfur dioxide (B) Carbon monoxide
(C) Carbon dioxide (D) Hydrocarbons.

2. One of the following is regarded as greenhouse gas
(A) Carbon monoxide (B) Carbon dioxide
(C) Sulfur dioxide (D) Nitrogen dioxide.

3. The most common inorganic polutant found in water, air, foods and soils
(A) Cadmium (B) Mercury
(C) Lead (D) Aluminium.

4. The following compound of kesari dal is implicated in lathyrism
(A) β-oxalylaminoalanine
(B) Cadmium (C) Aflatoxin
(D) Organophosphates

5. Ouch disease or itai itai is linked with the poisoning of
(A) Lead (B) Mercury
(C) Cadmium (D) Aluminium.

Answers

1. **(A)** 2.**(B)** 3. **(C)** 4.**(A)** 5. **(C)**.

For related clinical case study, refer 10.5

VIVA/Short Questions

1. What is metabolic heat?

Generation of heat by the body due to the ongoing biochemical processes.

2. What do you understand by heat stroke?

The failure of the heat regulatory system (thermoregulation) of the body.

3. What is environmental pollution?

Addition of extraneous (foreign) materials to air, water or land which adversely affects the quality of life.

4. What is ozone layer and what is its importance?

Elevation in CO_2 near earth's surface that traps sunlight and increases atmospheric temperature.

5. What is greenhouse effect?

Ozone is formed from atmospheric oxygen during high energy radiations of electrical discharges. This ozone forms a layer above the earth's surface (15-35 km). It absorbs harmful ultraviolet radiations of sun.

6. Name some water-borne diseases.

Typhoid, paratyphoid, cholera, amoebiasis, giardiasis and infectious hepatitis.

7. What is eutrophication?

This phenomenon of water deoxygenation is referred to as eutrophication.

8. What are the manifestations of lead toxicity in growing children?

Learning disabilities, behavioural changes (hyperexcitability) and mental retardation.

9. What are the effects of noise pollution?

Headache, increased blood pressure, irritability, neuromuscular tension, confusion, disturbed vision etc.

10. What is lathyrism?

Excessive consumption of kesari dal causes paralysis of lower limbs referred to as lathyrism.

11. What are goitrogens?

The compounds that prevent uptake and utilization of iodine by thyroid gland.

36. Insulin, Glucose Homeostasis, and Diabetes mellitus

MCQs

1. Increased blood glucose level (hyperglycemia) is an invariable feature of the disease

(A) Diabetes insipidus

(B) Diabetes mellitus

(C) Cushing's syndrome

(D) None of the above.

2. The following statement(s) is/are true about insulin. It was the first

(A) Hormone to be purified and synthesized

(B) Protein sequenced for amino acid composition

(C) Protein produced by the recombinant DNA technology

(D) All of the above are true.

3. Human insulin has two polypeptide chains namely A and B, respectively containing

(A) 21 and 30 amino acids

(B) 30 and 21 amino acids

(C) 19 and 32 amino acids

(D) 29 and 22 amino acids.

4. In human insulin structure, the two polypeptide chains are held together by disulfide bonds numbering

(A) 1 (B) 2 (C) 3 (D) 4.

5. Insulin is synthesized by the following cells of the pancreas

(A) α-cells (B) β-cells (C) γ-cells (D) δ-cells.

6. The polypeptide that serves as a useful index for the endogenous production of insulin

(A) A-peptide (B) B-peptide

(C) C-peptide (D) γ-peptide.

7. Insulin/proinsulin combines with the following metal to form complexes which are stored in the cytosomal granules

(A) Copper (B) Iodine

(C) Magnesium (D) Zinc.

8. The most predominant stimulus for insulin secretion

(A) A rise in blood glucose

(B) A fall in blood glucose

(C) A rise in plasma amino acids

(D) Gastrointestinal hormones.

9. The prominant inhibitor of insulin secretion

(A) Gastrointestinal hormones

(B) Epinephrine (C) Growth hormone

(D) Cortisol.

10. Insulin lowers blood glucose level by

(A) Promoting glucose uptake of tissues

(B) Increasing glycolysis in liver and muscle

(C) Decreasing gluconeogenesis

(D) All of the above.

11. Insulin is required for the uptake of glucose by the following tissues/organs except

(A) Cardiac muscle (B) Adipose tissue

(C) Leucocytes (D) Erythrocytes.

12. By stimulating the glucose utilization, insulin promotes the uptake of glucose by

(A) Brain (B) Liver

(C) Kidney (D) Erythrocytes.

13. Insulin increases the activities of the following enzymes except

(A) Glucokinase

(B) Pyruvate carboxylase

(C) Glycogen synthase

(D) Acetyl CoA carboxylase.

14. Insulin favours the synthesis of triacylglycerols from glucose by

(A) Providing more glycerol 3-phosphate

(B) Furnishing more NADPH via HMP shunt

(C) Increasing the activity of acetyl CoA carboxylase

(D) All of the above.

15. Insulin reduces ketogenesis by

(A) Decreasing the activity of HMG CoA synthase

(B) Promoting utilization of acetyl CoA via Krebs cycle

(C) Favouring lipogenesis

(D) All of the above.

16. The number of polypeptide subunits present in insulin receptor

(A) 1 (B) 2 (C) 3 (D) 4.

17. The phosphorylation of the following amino acid of the β-chain of the insulin receptor signals the biological actions of insulin

(A) Tyrosine (B) Threonine

(C) Serine

(D) Any one of the above.

18. The amino acid sequence of the following hormone is the same in all mammalian species so far studied

(A) Insulin (B) Glucagon

(C) Growth hormone (D) Prolactin.

19. The most potent hormone to elevate blood glucose level (hyperglycemia)

(A) Insulin (B) Glucagon

(C) Epinephrine (D) Thyroxine.

20. The following statements regarding the action of glucagon are true except it

(A) Increases glucose synthesis

(B) Enhances degradation of glycogen

(C) Acts through cyclic AMP

(D) Promotes glycolysis.

21. The pathway/cycle responsible for the conversion of muscle lactate to glucose

(A) Glycolysis (B) Cori cycle

(C) Krebs cycle (D) HMP shunt.

22. When the body is at rest (basal conditions), the blood glucose is predominantly utilized by

(A) Liver (B) Heart

(C) Kidney (D) Brain.

23. The renal threshold for glucose

(A) 120-160 mg/dl (B) 160-180 mg/dl

(C) 180-220 mg/dl (D) 220-240 mg/dl.

24. The renal tubular maximum for glucose (TmG)

(A) 80 mg/min (B) 180 mg/min

(C) 270 mg/min (D) 350 mg/min.

25. The blood glucose concentration is elevated by the action of the following hormones except

(A) Glucagon (B) Insulin

(C) Epinephrine (D) Thyroxine.

26. Insulin-dependent diabetes mellitus (type II diabetes) is characterized by

(A) Occurrence in the childhood (< 20 years age)

(B) Deficiency of insulin

(C) Ineffective oral hypoglycemic drugs

(D) All of the above.

27. Non-insulin dependent diabetes mellitus (type II diabetes) is characterized by the following except

(A) The prevalance is 80-90% of diabetes population

(B) Increase in the insulin receptors on the target cells

(C) Impairment in the production of insulin

(D) Effective treatment with oral hypoglycemic drugs.

28. The following precautions should be taken for performing oral glucose tolerance test except the subject should (be)

(A) Taking carbohydrate rich diet at least 3 days prior to the test

(B) On an overnight fast

(C) Do strenous exercise on the previous day of test

(D) Refrain from smoking and exercise during the test.

29. As per the WHO cirteria, a person is said to be suffering from diabetes if his/her fasting plasma glucose exceeds the following value

(A) 100 mg/dl (B) 126 mg/dl

(C) 140 mg/dl (D) 166 mg/dl.

30. Normally, glucose gets excreted into urine when the plasma glucose concentration exceeds

(A) 100 mg/dl (B) 140 mg/dl
(C) 180 mg/dl (D) 220 mg/dl.

31. The threshold for glucose is reduced in

(A) Diabetes mellitus (B) Renal glycosuria
(C) Alimentary glycosuria
(D) All of the above.

32. Severe uncontrolled diabetes mellitus is associated with the following metabolic complications

(A) Hyperglycemia (B) Ketoacidosis
(C) Hypertriglyceridemia
(D) All of the above.

33. The diabetic complications such as retinopathy, nephropathy and neuropathy are attributed to the accumulation of the following compound in the tissues/organs

(A) Glucose (B) Triacylglycerols
(C) Glycated proteins (D) Ketone bodies.

34. The normal glycosylated hemoglobin (HbA_{1c}) concentration, expressed as percentage of total hemoglobin

(A) 3-5% (B) 7-10%
(C) 12-15% (D) 17-20%.

35. The concentration of HbA_{1c} serves as an indication of the blood glucose concentration for the following period prior to its measurement

(A) 2-4 weeks (B) 6-8 weeks
(C) 10-12 weeks (D) 14-16 weeks.

36. Microalbuminuria refers to the excretion of albumin in urine in the range of

(A) < 30 mg/day (B) 30-300 mg/day
(C) 300-600 mg/day (D) None of the above.

Answers

1. **(B)** 2. **(D)** 3. **(A)** 4. **(B)** 5. **(B)** 6. **(C)** 7. **(D)** 8. **(A)** 9. **(B)** 10. **(D)** 11. **(D)** 12. **(B)** 13. **(B)** 14. **(D)** 15. **(D)** 16. **(D)** 17. **(A)** 18. **(B)** 19. **(B)** 20. **(D)** 21. **(B)** 22. **(D)** 23. **(B)** 24. **(D)** 25. **(B)** 26. **(D)** 27. **(B)** 28. **(C)** 29. **(B)** 30. **(C)** 31. **(B)** 32. **(D)** 33. **(C)** 34. **(A)** 35. **(B)** 36. **(B)**.

Clinical Case Study 36.1
(Source : Biochemistry, Pages 652-653)

Oral glucose tolerance test (mini GTT) was performed on a 45-year old obese woman with a family history of diabetes mellitus (her mother was a diabetic patient). The following are the laboratory data

Parameter	*Fasting*	*2 hr*
Blood glucose (mg/dl)	120	160
(Normal	<110	<140)
Urine glucose	–ve	–ve
Urine ketone bodies	–ve	–ve

1. **What is the probable diagnosis?**
2. **What is likely disorder she may develop? What are the precautions?**

Ans. :

1. The woman had marginally elevated fasting blood glucose and **impaired oral glucose tolerance test**.
2. Although she cannot be labelled as a diabetic patient at present, she requires careful periodical check up.

Clinical Case Study 36.2
(Source : Biochemistry, Page 651)

A 13-year old boy had the complaints of increased frequency of urination, and urinary tract infection. On routine examination, his urine was found to contain glucose and ketone bodies. He had a random blood glucose concentration 190 mg/dl. His laboratory data on the oral glucose tolerance test (mini GTT) are given below.

Parameter	*Fasting*	*2 hr*
Blood glucose (mg/dl)	150	240
(Normal	< 110	< 140)
Urine glucose	–ve	++
Ketone bodies	+	++

1. **What is the probable diagnosis? Justify.**
2. **What is the biochemical basis of this disorder?**
3. **What is the mode of treatment?**

Ans. :

1. The laboratory data of the boy-increase in blood glucose coupled with urinary excretion of glucose and ketone bodies, along with the results obtained in the OGTT point out that this is a good example of **juvenile onset diabetes mellitus** (type I diabetes or insulin dependent diabetes mellitus, IDDM).
2. IDDM, which mainly occurs in childhood, is characterized by almost total deficiency of insulin due to the destruction of β-cells of pancreas.
3. The boy has to be treated by insulin administration.

Clinical Case Study 36.3 (Source : Biochemistry, Pages 651-652)

A 45-year old woman visited her physician with complaints of increased appetite and thirst with high frequency of urination. She complained of generalized weakness, fatigue and sweating. Her laboratory findings on the oral glucose tolerance test (mini GTT) along with urine tests are as follows.

Parameter	*Fasting*	*2 hr*
Blood glucose (mg/dl)	155	205
(Normal Reference	<110	<140)
Urine glucose	–ve	++
Ketone bodies	–ve	–ve

1. **What is the probable diagnosis?**
2. **In which age group the disease mainly occurs?**
3. **What is the mode of treatment?**

Ans. :

1. The woman's fasting (elevated) blood glucose and the data obtained on OGTT indicate that she was suffering from **non-insulin dependent diabetes mellitus** (NIDDM or type II diabetes).
2. NIDDM mainly occurs in adults (>30 yrs) with a strong genetic predisposition.
3. The patients of NIDDM can be usually treated with oral antidiabetic drugs. Before prescribing the drugs, diet control and exercise are tried.

Clinical Case Study 36.4 (Source : Biochemistry, Page 653)

A healthy 25-year old pregnant woman (in 5th month of gestation), on routine examination, was found to contain glucose in her urine. Her random blood glucose was 150 mg/dl. She recalled that before pregnancy, her random blood glucose was <110 mg/dl and never she had glucose excreted into urine. Her mother was a diabetic patient. During her next visit a month later, her fasting and post-prandial blood glucose were respectively 130 mg/dl and 180 mg/dl, with traces of glucose in urine.

1. **What is the probable diagnosis?**
2. **What is the mode of treatment?**
3. **What is the future risk for this subject?**

Ans. :

1. **Gestational diabetes.**
2. Needs treatment with insulin during pregnancy.
3. This woman is likely to be a victim of diabetes in later years. Therefore, she has to be carefully monitored.

Clinical Case Study 36.5 (Source : Biochemistry, Page 655)

A 60-year old diabetic woman, who had been on insulin for the past 5 years, visited her physician for a routine check up. She informed that she was scrupulously following the dietary advise but not regular in insulin injection. The following are the laboratory data of this woman.

Parameter	*Subject*	*Reference range*
Blood glucose, fasting	100 mg/dl	70-100 mg/dl
Blood glucose, postprandial	140 mg/dl	130-160 mg/dl
Blood urea	25 mg/dl	15-40 mg/dl
HbA_{Ic} (% of total Hb)	9 %	3-5 %

Urine did not contain glucose or ketone bodies.

1. **What is the probable diagnosis?**
2. **What is the importance of estimating HbA_{1C} in diabeties?**

Ans. :

1. Elevated HbA_{Ic} (glycated hemoglobin) reflects **poor diabetic control** 6-8 weeks ago. This is frequently seen in diabetics who are irregular in the treatment regime.
2. The concentration of HbA_{Ic} serves as an indication of blood glucose level over a period equivalent to the half-life of RBC (hemoglobin) i.e., 6-8 weeks.

Clinical Case Study 36.5 (Source : Biochemistry, Page 656)

A 60-year old man, a patient of diabetes for 30 years had been on insulin for the past 5 years. His relevant laboratory data are given below.

Parameter	*Subject*	*Reference range*
HbA_{Ic} (% Hb)	5 %	3-5 %
Urine albumin	250 mg/day	<30 mg/day

1. **What is the probable diagnosis with justification?**
2. **Which organ function is likely to be adversely affected in this disorder?**

Ans. :

1. **Microalbuminuria**, which represents an intermediary stage between normal albumin excretion (< 30 mg/day) and macroalbuminuria (> 300 mg/day).
2. The small increase in albumin excretion predicts the onset of impairment in renal function in diabetic patients. Thus, microalbuminuria serves as a signal of early reversible renal damage (diabetic nephropathy).

Clinical Case Study 36.6
(Source : Biochemistry, Page 658)

A 65-year old man diabetic had been on oral antidiabetic drugs and for the past 7 years he had been on insulin. He had not been strictly following physician's instructions as regards his treatment schedule and dietary regime. He first developed microalbuminuria and now full fledged proteinuria. The following are his laboratory findings.

Parameter	*Subject*	*Reference range*
Blood glucose, fasting	220 mg/dl	70-110 mg/dl
Blood urea	90 mg/dl	15-40 mg/dl
Serum creatinine	4 mg/dl	1-1.5 mg/dl
Creatinine clearance	20 ml/min	120-145 ml/min

1. **What is the probable diagnosis with justification?**
2. **What is the likely mode of treatment?**

Ans. :

1. The diabetic patient was not on the proper treatment. Consequently, he had renal damage as is evident from the increased serum creatinine and drastically reduced creatinine clearance data, besides increase in blood urea level. This is a case of **diabetic nephropathy**.
2. The patient may require dialysis/renal replacement.

Clinical Case Study 36.7
(Source : Biochemistry, Pages 653-654)

A 30-year old man, on routine check up was found to excrete glucose in urine. His fasting and post-prandial blood glucose respectively were 90 mg/dl and 140 mg/dl. Oral glucose tolerance test (OGTT) also revealed that the man had a normal glucose tolerance except the presence of glucose in urine at 1 hr.

1. **What is the probable diagnosis?**
2. **What is the renal threshold for glucose?**
3. **Is this disorder related to diabetes?**
4. **What is the treatment?**

Ans. :

1. **Renal glycosuria**
2. Renal threshold for blood glucose (normal around 180 mg/dl).
3. Renal glycosuria is unrelated to diabetes mellitus and therefore not accompanied by the classical symptoms of diabetes.
4. This disorder does not require any specific treatment.

Clinical Case Study 36.8
(Source : Biochemistry, Page 653)

A 13-year old boy was constantly found to excrete glucose in urine after meals. A village doctor diagnosed as a case of type I diabetes and prescribed insulin administration. Following insulin injection, the boy had headache, sweating with anxiety, confusion and seizures. He became unconscious and was immediately shifted to the hospital. He was given intravenous glucose and the above symptoms disappeared.

1. **What is the probable diagnosis?**
2. **Why did the boy develop hypoglycemia?**

Ans. :

1. **Renal glycosuria, mistaken as diabetes mellitus.**
2. Without proper investigation (i.e., blood glucose, GTT). The boy was given insulin. He developed hypoglycemia symptoms.

Clinical Case Study 36.9
(Source : Biochemistry, Page 654)

A 35-year old man, with no family history of diabetes was found to excrete glucose in urine following a meal. His fasting glucose was found to be within normal limits-95 mg/dl. The following are the results of oral glucose tolerance test (OGTT).

Parameter	*Fasting*	*1 hr*	*2 hr*
Blood glucose (mg/dl)	95	190	135
Urine glucose	–ve	+	–ve

1. **What is the probable diagnosis with justification?**
2. **What is likely cause of this disorder?**

Ans. :

1. The subject had a transitory rise in blood glucose, following a meal. And this concentration exceeds renal threshold for glucose, it gets excreted in urine. This is a case of **alimentary (lag storage) glycosuria**.
2. Due to increased rate of intestinal absorption of glucose into blood stream.

For more clinical case studies, refer 13.4, 14.3, 19.1, 19.5, 21.1 and 21.2

VIVA/Short Questions

1. **What is diabetes mellitus?**

Diabetes mellitus is a clinical condition characterized by increased blood glucose level (hyperglycemia) due to insufficient or inefficient (incompetent) insulin.

2. What do you know about the structure of insulin and how is it synthesized?

Human insulin (mol wt. 5,734) contains 51 amino acids, arranged in two polypeptides chains. A chain has 21 amino acids while B has 30 amino acids, both are held together by two interchain disulfide bonds. Insulin is produced by the β-cells of the islets of Langerhans of pancreas.

3. What is a C-peptide?

Proinsulin on degradation gives equimolar concentration of connecting peptide (C-peptide) and insulin. C-peptide has no biological activity, however, its estimation in the plasma serves as a useful index for the endogenous production of insulin.

4. What factors influence insulin secretion?

Glucose, amino acids and gastrointestinal hormones stimulate insulin secretion Growth hormone, cortisol and estrogens increase insulin secretion while epinephrine inhibits.

5. What are the effects of insulin on carbohydrate metabolism?

Insulin lowers blood glucose level (hypoglycemic effect) by promoting its utilization and storage, and by inhibiting its production. Insulin is required for glucose uptake by some tissues (muscle, adipose tissue, mammary gland etc.). Further, insulin increases glycolysis and decreases gluconeogenesis.

6. How is lipid metabolism influenced by insulin?

Insulin favours the synthesis of triacylglycerols (lipogenesis) and decreases lipolysis, besides reducing the synthesis of ketone bodies.

7. What are the effects of insulin on protein metabolism?

Insulin stimulates the entry of amino acids into the cells, enhances protein synthesis and reduces protein degradation.

8. What is insulin receptor?

Insulin receptor is a tetrameric protein consisting of 4 subunits of 2 types and is designed as $\alpha_2\beta_2$. The insulin receptor is synthesized by a single gene.

9. How does insulin — mediated glucose transport occur?

The binding of insulin to insulin receptors signals the translocation of vesicles containing glucose transporters from intracellular pool to the plasma membrane. The glucose transporters are responsible for the insulin — mediated uptake of glucose by the cells.

10. What is glucagon and how is its synthesis regulated?

Glucagon is a polypeptide (with 29 amino acids) hormone synthesized by α-cells of the pancreas.

11. What are the metabolic effects of glucagon?

Glucagon influences carbohydrate, lipid and protein metabolisms. In general, the effects of this hormone oppose that of insulin. Glucagon increases blood glucose concentration (hyperglycemic), promotes fatty acid oxidation and increases the amino acid uptake by the liver.

12. Explain the terms hyperglycemia, hypoglycemia and glycosuria?

Hyperglycemia refers to an increase in the blood glucose above the normal level (70-100 mg/dl). Hypoglycemia represents decreased blood glucose concentration. Excretion of glucose into urine is known as glycosuria.

13. What are the sources of blood glucose?

The dietary carbohydrates, synthesis of glucose from non-carbohydrate sources (gluconeogenesis), degradation of glycogen (glycogenolysis) are the major contributors to blood glucose.

14. What are the major routes for the utilization of glucose?

Glycolysis, glycogenesis, HMP shunt and lipogenesis.

15. What is the role of kidney in the homeostasis of blood glucose?

Glucose is continuously filtered by the glomeruli, reabsorbed and returned to the blood. If the level of glucose in blood is above 160-180 mg/dl, it is excreted in urine (glycosuria). This value is referred to as renal threshold for glucose.

16. What is tubular maximum for glucose?

The maximum ability of the renal tubules to reabsorb glucose per minute is known as tubular maximum for glucose (TmG). The value for TmG is 350 mg/min.

17. What is the role of hormones in blood glucose homeostasis?

Insulin lowers blood glucose (hypoglycemic) while the rest of the hormones (glucagon, epinephrine, thyroxine, glucocorticoids, growth hormone etc.) oppose the action of insulin. Thus, these hormones (except insulin) act in such a way that the blood glucose level is elevated (hyperglycemic).

18. What do you know about hypoglycemia?

A reduction in blood glucose level (normal 70-100 mg/dl) is referred to as hypoglycemia. The manifestations of hypoglycemia (headache, anxiety, confusion, sweating, slurred speech) are observed when the blood glucose concentration falls below 45 mg/dl.

19. How is diabetes mellitus classified?

Insulin dependent diabetes mellitus (IDDM, type I diabetes, juvenile onset diabetes) and non-insulin dependent diabetes mellitus (types II diabetes, adult onset diabetes). This classification is mainly based on the requirement of insulin for treatment.

20. What do you know about non-insulin dependent mellitus (NIDDM)?

NIDDM is more common, accounting for 80-90% of the diabetes populaton. It occurs in adults (usually above 35 years). Thé causative factors for NIDDM include genetic and environmental.

21. What is glucose tolerance test (GTT)?

The individual's response to oral glucose load is referred to as glucose tolerance test.

22. What is the WHO criteria for the diagnosis of diabetes mellitus by oral GTT?

As per WHO norms, a person is said to be suffering from diabetes if his/her plasma glucose exceeds 126 mg/dl at fasting and 200 mg/dl at 2 hrs, after GTT.

23. What do you know about renal glycosuria?

Renal glycosuria is a benign condition due to a reduced renal threshold for glucose (normal 160-180 mg/dl).

24. What are the major metabolic changes associated with diabetes?

Hyperglycemia, ketoacidosis and hypertriglyceridemia.

25. What is glycosylated hemoglobin?

Glycated or glycosylated hemoglobin refers to the glucose derived products of normal adult hemoglobin (HbA). Among the glycated hemoglobins, the most abundant form is HbA_{1c}. It is produced by the condensation of glucose with N-terminal valine of each β-chain of HbA.

26. What is microalbuminuria and what is its diagnostic importance?

Microalbuminuria refers to the excretion of 30–300 mg of albumin in urine per day. The small increase in albumin excretion predicts impairment in renal function in diabetic patients.

37. Cancer

MCQs

1. **Cancer is derived from a Latin word meaning**
(A) Web (B) Scorpion (C) Crab (D) Snake.
2. **Cancer holds the following rank in the order of killer diseases in the developed countries**
(A) 1st (B) 2nd (C) 3rd (D) 4th.
3. **More than 70% of the cancers occur in persons in the age of**
(A) 10-20 years (B) 20-40 years
(C) 40-60 years (D) 60 years and above.
4. **The entry of chemical carcinogens into the body occurs through**
(A) Diet (B) Drugs
(C) Environmental pollutants
(D) Any one of the above.
5. **Ames assay is a laboratory test to check the carcinogenecity of chemicals. It employs the following organism**
(A) *Escherichia coli*
(B) *Haemophilus aegypticus*
(C) *Salmonella typhimurium*
(D) Any one of the above.
6. **The genetic material in oncogenic viruses**
(A) DNA only (B) DNA or RNA
(C) RNA only (D) Protein only.
7. **The genes that regulate the development of cancer**
(A) Oncogenes (B) Antioncogenes
(C) Apoptosis genes (D) All of the above.
8. **Burkitt's lymphoma, a cancer of B-lymphocytes is a good example of following category of activation of protooncogenes to oncogenes**
(A) Viral insertion into chromosome
(B) Gene amplification
(C) Chromosomal translocation
(D) Point mutation.
9. **The following is/are the oncoprotein growth factor(s) synthesized by the oncogenes**
(A) Epidermal growth factor
(B) Erythropoietin
(C) Tumor necrosis factor
(D) All of the above.
10. **Apoptosis means programmed cell**
(A) Growth (B) Differentiation
(C) Development (D) Death.

11. The tumor marker cancer antigen-125 (CA-125) is used for the detection of
(A) Prostatic cancer (B) Multiple myeloma
(C) Ovarian cancer (D) Cancer of colon.

12. The following are the morphological changes in the growing tumor cells except they
(A) Are much rounder in shape
(B) Lack contact inhibition
(C) Possess unaltered membrane permeability
(D) Have altered cytoskeleton structure.

13. The growing tumor cells are characterized by the following biochemical character(s)
(A) Increased replication and transcription
(B) Enhanced glycolysis
(C) Reduced requirement of growth factors
(D) All of the above.

Answers

1. **(C)** 2. **(B)** 3. **(D)** 4. **(D)** 5. **(C)** 6. **(B)** 7. **(D)** 8. **(C)** 9. **(D)** 10. **(D)** 11. **(C)** 12. **(C)** 13. **(D)**.

Clinical Case Study 37.1 (Source : Biochemistry, Page 663)

A 68 year old woman had complaints of bloating, abdominal discomfort, post-menopausal vaginal bleeding, loss of appetite and fatigue. Her pelvic examination and ultrasound indicated ovarian mass. Her laboratory data are given below.

Parameter	*Subject*	*Reference range*
CA-125	280 U/ml	0-35 U/ml

1. **What is the probable diagnosis?**
2. **What are the additional tests needed to confirm this disorder?**
3. **Write briefly on tumor markers.**

Ans. :

1. The old woman with an elevation in serum CA-125 levels, and pelvic mass appears to be a victim of **ovarian cancer**.
2. Laparotomy and histopathology.
3. Refer Chapter 31, pages 662-663.

Clinical Case Study 37.2 (Source : Biochemistry, Page 663)

A 60-year old woman complained of back pain, non-responsive to drugs. The physical examination revealed that a small marble-sized painless mass in the neck, with an existence of 15 years, had started increasing in size during the recent 4 months. Serum electrophoresis showed a distinct band of monoclonal immunoglobin between β- and γ-globulins. The urine gave a positive test for Bence Jones proteins.

1. **What is the probable diagnosis, with justification?**
2. **What is the biochemical basis of this disorder?**
3. **What is the immunological status of these patients?**

Ans. :

1. The characteristic electrophoretic pattern of serum proteins with an 'M' band along with the excretion of Bence Jones proteins in urine and the clinical manifestations led to the diagnosis that the patient was suffering from **multiple myeloma**.
2. Malignancy of a single clone of plasma cells in the bone marrow resulting in an abnormal production of immunoglobulins, mostly IgG (75%) and in some cases IgA (25%).
3. The patients of multiple myeloma have depressed immunity, hence recurrent infections are common in them.

For more clinical case studies, refer 6.4 and 19.6

VIVA/Short Questions

1. What do you understand by the term cancer?

Cancer is characterized by loss of control of cellular growth and development leading to excessive proliferation and spread of cells.

2. What is metastasis?

The process of proliferation and spread of cells to various parts of the body.

3. What are carcinomas and sarcomas?

Carcinomas refer to the cancers of epithelial cells while sarcomas refer to the cancers of connective tissues.

4. What are the causative agents of cancer?

In general, cancers are multifactorial in origin. The causative agents include physical, chemical, genetic and environmental factors.

5. What do you know about chemical carcinogens?

It is estimated that almost 80% of human cancers are caused by chemicals which may be organic (dimethyl benzanthracene) or inorganic (arsenic) in nature. The entry of chemical carcinogens into the body may occur through diet and drugs, besides the occupation and life styles.

6. What is probable mechanism of action of carcinogens?

The carcinogens can covalently bind to purines, pyrimidines and phosphodiester bonds of DNA, often causing unrepairable damage. The chemical carcinogens frequently cause mutations which may finally lead to the development of cancer, hence they are regarded as chemical carcinogens.

7. What is Ames assay?

Ames assay is the laboratory test to check the carcinogenecity of chemicals. It is carried out by employing a mutant strain of bacterium, namely *Salmonella typhimurium.*

8. What do you know about carcinogenic viruses?

The viruses involved in the development of cancers are commonly known as carcinogenic viruses or oncogenic viruses. They may contain DNA (e.g., adenoviruses) or RNA (e.g., retrovirus type B) as the genetic material.

9. What are oncogenes?

The genes capable of causing cancers are known as oncogenes.

10. What are protooncogenes?

The genes of the normal host cells which are similar to viral oncogenes are referred to as proto-oncogenes.

11. How are protooncogenes acivated to oncogenes?

Viral insertion into chromosome, chromosomal translocation, gene amplification and point mutation are some of the important mechanisms for converting protooncogenes to oncogenes.

12. What is the mechanism of action of oncogenes?

Oncogenes encode for certain proteins namely oncoproteins (growth factors, growth factor receptors, GTP-binding proteins). These proteins are the altered versions of their normal counterparts and are involved in the transformation and multiplication of cells.

13. What do you know about antioncogenes?

Antioncogenes or cancer suppressor genes result in the production of certain products which apply breaks and regulate cell proliferation. A defect in these genes removes growth control of cells and is believed to be a key factor in the development of several tumours.

14. What is apoptosis?

Apoptosis is the phenomenon of programmed cell death.

15. What are tumour markers?

The biochemical indicators employed to detect the presence of cancers are collectively referred to as tumor markers. e.g., carcinoembryonic antigen (for cancers of colon, stomach, lung), cancer antigen — 125 (for ovarian cancer), prostate specific antigen (for prostate cancer).

16. What are the biochemical characteristics of growing tumour cells?

The tumour cells are characterized by increased replication and transcription, reduced requirement of growth factors, increased glycolysis, enhanced synthesis of fetal proteins etc.

38. Acquired Immunodeficiency Syndrome (AIDS)

MCQs

1. AIDS was first reported in 1981 in

(A) Intravenous drug abusers
(B) Homosexuals (C) Heterosexuals
(D) Blood transfused subjects.

2. Transmission of HIV occurs most predominantly in

(A) Homosexuals (B) Heterosexuals
(C) Drug abusers (D) Pregnant women.

3. The genetic material present in human immunodeficiency virus

(A) DNA (B) DNA and RNA
(C) RNA (D) Protein.

4. The HIV genome contains 3 structural genes-*gag, pol,* and *env* respectively coding for

(A) Core proteins, reverse transcriptase and envelop protein
(B) Reverse transcriptase, core protein and envelop protein
(C) Envelop protein, core protein and reverse transcriptase
(D) Core protein, envelop protein and reverse transcriptase.

5. Most of the immunodeficiency symptoms of AIDS are associated with

(A) B-lymphocytes (B) T-lymphocytes
(C) CD_4-cells (D) Complement system.

6. Seropositive period in AIDS is observed in

(A) Acute phase (B) Chronic phase
(C) Crisis phase (D) None of the above.

7. For the diagnosis of AIDS, the CD_4 T-lymphocytes in blood must fall to

(A) < 800/μl (B) < 600/μl
(C) < 400/μl (D) < 200/μl.

8. The HIV genome can be identified by the following test
(A) Enzyme-linked immunosorbant assay
(B) Western blot technique
(C) Polymerase chain reaction
(D) Any one of the above.

9. The following two drugs are commonly employed for the treatment of AIDS to prolong life
(A) Zidovudine and methotrexate
(B) Zidovudine and didanosine
(C) Zidavudine and azaserine
(D) Didanosine and azaserine.

10. Zidovudine is used for the treatment of AIDs. This drug inhibits
(A) Reverse transcriptase (B) DNA helicase
(C) RNA polymerase (D) Cyclooxygenase.

Answers

1. **(B)** 2. **(A)** 3. **(C)** 4. **(A)** 5. **(C)** 6. **(B)** 7. **(D)** 8. **(C)** 9. **(B)** 10. **(A)**.

Clinical Case Study 38.1
(Source : Biochemistry, Pages 668-670)

A 30-year old man had fever, diarrhea, weight loss. He had multiple opportunistic infections, generalized lymphadenopathy, besides neurological complications. He had 150 CD_4T-lymphocytes/μl blood (normal >800/ μl). (ELISA) was positive for the antibodies of human immunodeficiency virus (HIV).

1. **What is the probable diagnosis with justification?**
2. **What is the status of immunity in these patients?**
3. **Write briefly on the drugs employed for the treatment of this disease.**

Ans. :

1. The presence of HIV antibodies and the reduced concentration of CD_4T-lymphocytes in the circulation clearly indicate that this is a case of **AIDS**.
2. Due to deficiency of the immune system, the body of AIDS patient is highly susceptible to all sorts of infections (viral, bacterial, fungal). This is in contrast to healthy individuals.
3. Refer Chapter 38, page 670.

VIVA/Short Questions

1. What are the major characteristic features of AIDS?

AIDS is a retroviral disease, caused by human immunodeficiency virus (HIV). It is characterized by immunosuppression, secondary neoplasma and neurological manifestations. AIDS is invariably fatal, since there is no cure.

2. How is the HIV transmitted?

Transmission of HIV (i.e., AIDS) essentially requires the exchange of body fluids (semen, vaginal secretions, blood, milk) containing the virus or virus-infected cells. Sexual intercourse is the major risk factor in AIDS transmission (60% in homosexuals and 15% in sex between men and women).

3. What are the immunological abnormalities in AIDS?

As the name indicates, immunodeficiency (or immunosuppression) is the hallmark of the disease. AIDS primarily affects the cell-mediated immune system which protects the body from intracellular parasites such as viruses, protozoa and mycobacteria. This is caused by a reduction in CD_4 cells of T-lymphocytes.

4. What is the importance of CD_4 cells?

CD_4 (cluster determinant antigen 4) cells of T-lymphocytes are the master cells of cell mediated immunity. Most of the immunodeficiency symptoms of AIDS are associated with the reduction in CD_4 cells.

5. What is the natural course of AIDS in humans?

Three distinct phases of HIV interaction with the immune system of infected body have been identified. There are the early, acute phase;, the intermediate, chronic phase; the final, crisis phase.

6. What are the laboratory tests employed to diagnose HIV infection?

Enzyme-linked immunosorbant assay (ELISA) and Western blot technique for the detection of AIDS. Polymerase chain reaction (PCR) has been recently used for the detecton of HIV genome.

7. What are the drugs used for the treatment of AIDS?

Zidovudine or 3′ azido 2,3′-dideoxythymidine (AZT) and didanosine are the two drugs employed to treat AIDS.

8. Why has it not been possible to develop vaccine against AIDS?

HIV exhibits genetic heterogenecity with a result that several species of the virus may be found in the same patient. The principal causes for genetic variation is due to lack of proof-reading activity by the enzyme reverse transcriptase. This leads to very frequent alteration in the DNA base sequence and consequently the amino acid sequence of proteins. For this reason, it has not been possible to develop a vaccine against AIDS.